BASIC ANATOMY

THIRD EDITION

BASIC ANATOMY

A LABORATORY MANUAL

The Human Skeleton • The Cat

B. L. Allen

Ohio University

W. H. FREEMAN AND COMPANY

New York

ISBN 0-7167-1755-7

Printed in the United States of America

Twelfth printing, 1998

CONTENTS

ILLUSTRATIONS

FIGURES

HUMAN SKELETAL DIAGRAMS

PREFACE

In many elementary anatomy courses, such as those for students in physical education, pre-physical therapy, prenursing, premortuary, and medical technology, it is desirable to emphasize the structure of the human body. Seeing and handling the actual structures is very helpful for learning anatomy and understanding anatomical relationships. However, in most undergraduate colleges the use of human cadavers for dissection is not feasible, because it is difficult to obtain them and special facilities are needed for their storage. Instead the cat is often used, since it is anatomically very similar to the human, and it is less expensive and easier to store. But for study of the skeletal system, the human skeleton and the separate bones are easily obtained and stored, and the cost is not prohibitive. This laboratory manual, then, has been designed for courses in which the human skeleton and the cat are studied. Where there are pertinent differences between the cat and the human, they are noted; the students should know these differences, and make use of their own living specimens (themselves) as much as possible, to help them learn and appreciate the structure of the human body, as well as that of the cat.

The material is organized systemically, but some relevant regional features are included with the study of the muscles of the appendages. The directions for the study of each system are more detailed than will be necessary for every class, and it is assumed that instructors will omit the material that they consider nonessential for particular courses.

To aid in the identification of various structures, there are numerous illustrations of the skull and other components of the human skeleton, as well as an illustration of the articulated human skeleton and one of the cat skeleton. There are in addition many illustrations of the musculature and the internal organ systems of the cat. A list of definitions for special features, or markings, on bones is included for students who are studying bones in detail; it is intended as a reference, rather than as a set of terms to be memorized. Students who are particularly concerned with human muscles can draw these in place on the human skeletal diagrams provided at the end of the chapter on the muscular system.

The illustrations and the directions for study and dissection have been designed to enable the students to progress in their laboratory work with little help from the instructor. The students should always try to locate, externally on themselves, as many structures as they can, and to be aware that the structures being studied are not isolated, but in life are functionally integrated parts within a system, which in turn is functionally integrated with the other systems to form a whole working body.

In this third edition of the manual, following the suggestions of various instructors, a brief discussion of joints has been added to the chapter on the skeletal system and additional actions of muscles and their nerve supplies are given in the chapter on the muscular system. The chapters on the skeletal, circulatory, urogenital, and nervous systems include new illustrations which should prove useful for study of the skeleton and in dissection procedures.

ACKNOWLEDGMENTS

I wish to express my sincere appreciation to a number of people for their help in preparing the first and second and now this third edition of the manual. First and foremost are my students, who suggested that I should prepare a manual for them. Others are: Rush Elliott, the late emeritus professor of anatomy of Ohio University, who read the manuscript for the first edition and made many valuable suggestions; Doreen Davis Masterson, who did dissection, layouts, and final drawings for the first edition; Edward Hanson, Julia P. Iltis, Edna Indritz-Steadman, Jill Leland, Margaret L. Muller, and Tom Moore, who also prepared final drawings; and James A. Dodd, Nancy Field, Friedrich Schulte, and Stephen Wagley at W. H. Freeman and Company. I would also like to offer special thanks to the following, whose comments and suggestions were very helpful to me as I worked on the new edition: Richard E. Jones, University of Colorado, Boulder; Dan F. Penney, San Jacinto College; and G. E. Thompson, El Camino College. I am also sincerely grateful to and wish to express my thanks to the following: the educators and students who have given me encouragement by using the manual; the publisher, whose patience is highly commendable; and all of the others who have had a part in preparing this revision.

April 1986   **B. L. Allen**

INTRODUCTION TO THE STUDENT: GENERAL DIRECTIONS

USE OF LABORATORY MATERIALS

At all times, handle with care the fragile bone specimens or other materials that may be provided for your use. When it is necessary to share specimens with others, be unselfish and courteous.

PRELIMINARY STUDY

Become familiar with directional terms and planes of section that are used with reference to the body. You should also have some knowledge of the kinds of tissues that make up the organs of the body.

Directional Terms
(Fig. 0-1)

Cranial, or **cephalic** The head end. (**Craniad,** or **cephalad** Toward the head.)

Caudal The tail end. (**Caudad** Toward the tail.)

Ventral The belly side. (**Ventrad** Toward the belly.)*

Dorsal The back side. (**Dorsad** Toward the back.)*

*In reference to the hindlimb, the terms "dorsal" and "ventral" differ between the cat and the human. In the cat the anterior aspect of the hindlimb is dorsal and the posterior aspect is ventral. In the human "dorsal" refers to the posterior aspect of the hindlimb, and "ventral" to the anterior. In this manual the terms corresponding to dorsal and ventral in the human will be used throughout to avoid confusion.

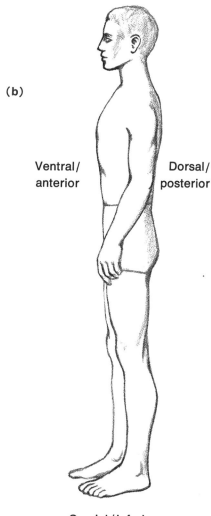

Cranial/superior

(b)

Ventral/
anterior

Dorsal/
posterior

Caudal/inferior

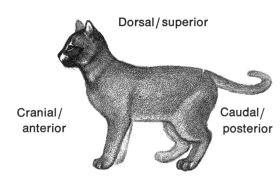

(a)

Dorsal/superior

Cranial/
anterior

Caudal/
posterior

Ventral/inferior

Figure 0-1
Directional terms

(a) Quadruped
(b) Biped

Superior The upper portion, or that which is above.

Inferior The lower portion, or that which is below.

Anterior Used synonymously with "cranial" in quadrupeds, and with "ventral" in bipeds.

Posterior Used synonymously with "caudal" in quadrupeds, and with "dorsal" in bipeds.

Median The middle, or the midline.

Midline An imaginary plane that bisects the body into right and left halves.

Medial Refers to a location nearer to the midline than another location. (**Mediad** Toward the midline.)

Lateral Refers to a location farther from the midline than another location. (**Laterad** Away from the midline.)

Proximal Commonly used with reference to the appendages, meaning that portion which is nearer the trunk or main body mass (nearest to point of attachment to the trunk). (**Proximad** Toward the trunk.)

Distal Used in conjunction with proximal, meaning farther away from the trunk (farthest from point of attachment). (**Distad** Away from the trunk.)

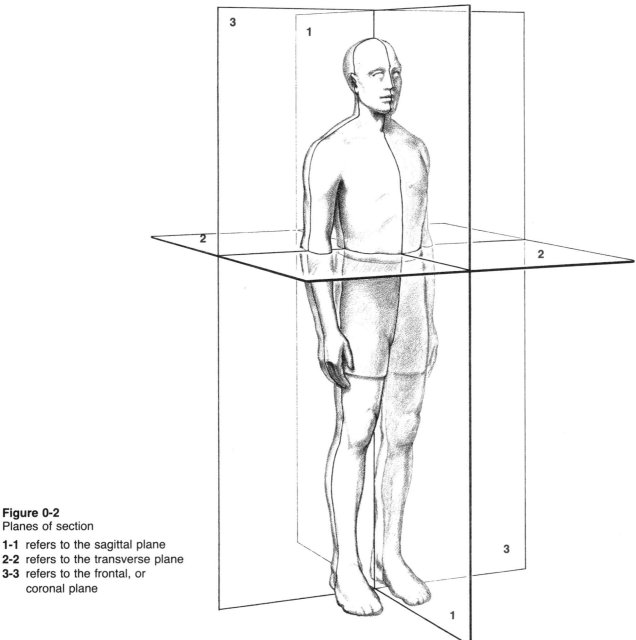

Figure 0-2
Planes of section

1-1 refers to the sagittal plane
2-2 refers to the transverse plane
3-3 refers to the frontal, or
    coronal plane

Planes of Section
(Fig. 0-2)

Transverse, or **cross** A plane that extends from left to right and from dorsal to ventral, giving cranial and caudal portions.

Longitudinal A plane that extends from cranial to caudal; longitudinal and transverse planes intersect at right angles.

Sagittal, or **midsagittal** A longitudinal plane that passes through the midline to divide the body into right and left halves.

Parasagittal A longitudinal plane that parallels the sagittal plane to either the right or the left.

Frontal, or **coronal** A longitudinal plane that extends from left to right and cranial to caudal, giving dorsal and ventral portions.

Oblique, or **diagonal** A plane that is neither cross nor longitudinal, but somewhere between the two. One frequently obtains this type of plane whether intended or not.

Tissues of the Body

You should know from the study of general biology that all living things are composed of cells and cell products. When examining the human, or the cat, you should be aware that groups of similar cells and their products form tissues, which are arranged in various ways to form organs; organs function together to form an organ system, and these organ systems are functionally integrated to produce a living organism.

The tissues that form organs are of four primary kinds: (1) **epithelial,** which covers body surfaces and lines hollow organs and cavities; (2) **connective,** a binding and supporting tissue; (3) **muscular,** which is specialized for contraction and thus responsible for movement; and (4) **nervous,** which is specialized for receiving stimuli and conducting impulses. Within each primary category specializations occur that adapt the tissues for the specific function(s) they are to perform at a given location.

You should consult your textbook, or a textbook of histology, for detailed accounts and illustrations to renew your knowledge of the tissues that make up the specimens you will use in the laboratory.

BASIC ANATOMY

1 HUMAN SKELETAL SYSTEM

Observe skeletons of both the human and the cat, and make comparisons. It is important to become oriented with the cat skeleton as well as with the human, since the cat will be used for dissection. Use the illustrations and descriptions that are included in this manual as study aids in identifying structures and various details indicated in the outline that follows.

The important joints should be studied in conjunction with the various bones that are involved in these joints, and also the movements that are possible in the joints. The following discussion of joints is intended to help you. Also, a list of basic bone features is given on page 6 for classes in which details of individual bones are studied.

ARTHROSES, OR JOINTS*
(Fig. 1-1)

Arthroses, or *joints* in common parlance, are not usually dissected in the cat. However, it will be useful to know what joints are present between bones and what movement can be produced by contraction of muscles crossing them. A muscle, or its tendon, may cross one or several joints and have an action on each joint it crosses. Ways of classifying joints and the types of movement they produce are given below.

*Also commonly called *articulations*.

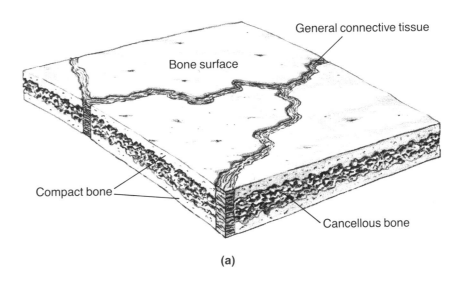

(a)

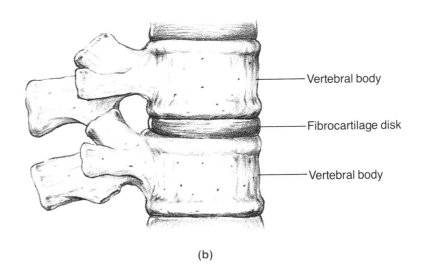

(b)

Classification of Joints

One way of classifying joints is by structure, ie, by the particular kinds and arrangement of tissues between the bones of a joint. Another classification is by the amount of movement, if any, permitted within the joint. A third way is by the number of planes in which movement is produced.

Structure

Fibrous The bones held together by a general type of connective tissue. An example is the *suture* joint that occurs between most bones of the skull. A wider fibrous connection having more flexibility is called a *syndesmosis*; an example is the interosseous membrane between the shafts of the leg bones. The interosseous membrane between the forearm bones is wide enough to permit the radius to rotate around the ulna.

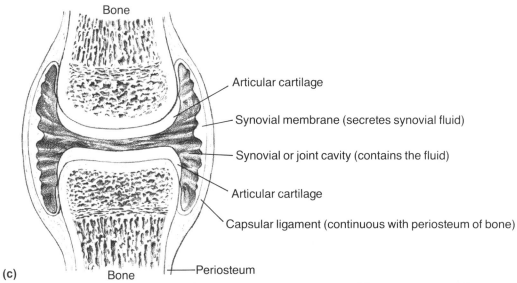

Bone

Articular cartilage

Synovial membrane (secretes synovial fluid)

Synovial or joint cavity (contains the fluid)

Articular cartilage

Capsular ligament (continuous with periosteum of bone)

Periosteum

(c) Bone

Figure 1-1

(a) Fibrous joint, showing suture joints found in the skull
(b) Cartilaginous joint, showing a symphysis joint between bodies of two vertebrae
(c) Synovial joint *A*, showing a longitudinal section between the ends of two long bones
(d) Synovial joint *B*, showing a joint cavity divided by an intra-articular disk. Such a disk is usually composed of fibrocartilage.

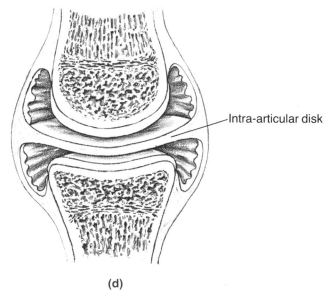

Intra-articular disk

(d)

Cartilaginous A cartilage disk or plate between the bones. Most of these are fibrocartilage joints, such as the *symphysis* joints between the bodies of vertebrae.

Synovial A joint cavity and other distinctive features present between the bones (see Fig. 1-1, c and d).

Amount of movement permitted

Synarthrosis No true movement is permitted, as in a suture joint.

Amphiarthrosis Limited movement is permitted. In these cases, however the cartilage is merely compressed and the wide fibrous connections flexed rather than an actual movement occurring between bone surfaces.

Diarthrosis "Free" movement is permitted, although the movement is only as free as the configuration of opposing articular surfaces and the strength and placement of accessory ligaments allow. Diarthroses are synovial joints and the movements they produce are:

Flexion Decrease in the size of an angle between bones, as when one bends the elbow.

Extension Increase in the size of an angle, as when one straightens the bent elbow.

Hyperextension A reverse flexion, as when swinging the arm dorsad.

Abduction Movement away from a midline plane, as when moving the arm out to the side and up.

Adduction Movement that returns the abducted part toward a midline plane.

Circumduction Movement that circumscribes a cone, as when swinging the arm around in a circle. This is a combination of the movements given above.

Rotation A bone may rotate around its own longitudinal axis in an inward direction, *medial rotation*, or in an outward direction, *lateral rotation*. These movements occur in both the shoulder and hip joints. In another type of rotation, sometimes called *"true" rotation*, one bone remains stationary and another bone rotates partially around it, as when the radius of the forearm rotates around the ulna. In the forearm, rotation is referred to as *supination* when the palm of the hand is turned ventrad and as *pronation* when the palm is turned dorsad.

Gliding This, the simplest movement between bones, occurs when one relatively flat surface glides over another such surface. A movement of the foot, in which the sole is directed laterad, *eversion*, and the opposite movement, *inversion*, in which the sole is directed mediad, are both produced by gliding movements between the tarsal bones.

Number of planes in which movement occurs

Uniaxial Movement occurs in one plane only, as in the elbow joint.

Biaxial Movement occurs in two planes, as in the wrist joint.

Multiaxial Movement occurs in three or more planes, as in the shoulder joint.

Diarthroses, or synovial joints, are further categorized as follows:

Hinge, or **ginglymus** A uniaxial joint permitting only flexion and extension, and hyperextension in some cases. The ankle joint is an example.

Condyloid A biaxial joint that permits flexion, extension, perhaps hyperextension, and abduction and adduction. A condyloid joint has a rounded convexity that articulates with a complementary concavity. Examples are the knuckles of the hand (the metacarpophalangeal joints, not the hinge-type interphalangeal joints).

Ellipsoid Like the condyloid joint but more shallow, as in the wrist.

Rotary, pivot, or **trochoid** A uniaxial joint in which one bone remains stationary while another rotates around it. The movement permitted is "true" rotation. Two such joints occur between the radius and ulna of the forearm, one at the proximal ends and one at the distal ends.

Plane, gliding, or **arthrodial** This type of joint has relatively flat articular surfaces, or just slightly convex and concave ones. It is multiaxial and permits gliding movements in various directions. The joint between the collar bone and shoulder blade is just one example of the numerous gliding joints in the body.

Ball-and-socket, or **enarthrodial** A pronounced convexity, called a head, articulates with a concavity, as in the shoulder and hip joints. These multiaxial joints permit all movements except gliding and true rotation.

Saddle, or **sellar** One articular surface is saddle-shaped and the opposing surface is its converse. This multiaxial joint permits all movements except gliding and rotation. The joint at the proximal end of the hand bone (first metacarpal) that leads to the thumb is a saddle joint.

Consult your textbook for descriptions and illustrations of individual joints. **Caution:** Do *not* use the articulated skeletons for producing joint movements. *Use your own flexible joints*—they are not held together by wires as they are in the articulated skeleton.

DIVISIONS OF THE SKELETON
(Figs. 1-2, 1-3)

Axial Division

Skull

Cranium The portion of the skull that encases the brain.
Face Region of the forehead, eyes, nose, cheeks, and jaws.
Ear ossicles These three small bones are enclosed in the petrous portion of the temporal bone and will not be seen.
Hyoid A single bone having no actual bony connections with the skull.

Vertebral column The following regions are recognized:

Cervical The neck.
Thoracic The "chest."
Lumbar The "small" of the back.
Sacral The pelvic area.
Coccygeal, or **caudal** The tail.

Thorax

Sternum The "breastbone."
Ribs The bones connecting the thoracic vertebrae and the sternum; all ribs together form the rib cage.

Appendicular Division

Pectoral appendages The cranial or superior appendages.

Pectoral girdle(s) A partial girdle on each side, consisting of a shoulder blade (scapula) and a collar bone (clavicle). Some anatomists refer to a single girdle, completed ventrally by the sternum, but it is incomplete dorsally.
Pectoral extremities The arm, forearm, and hand on each side.

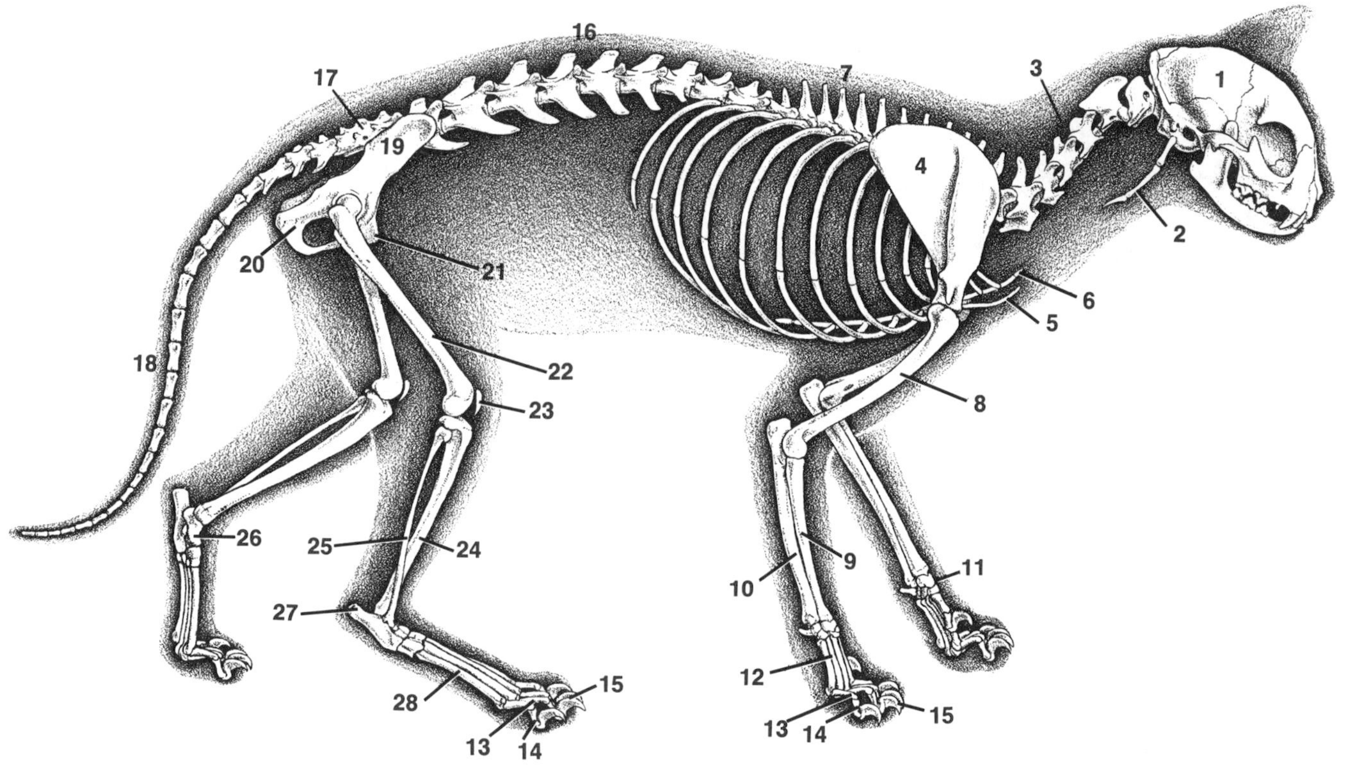

Figure 1-2
Cat skeleton, lateral view

  1 Skull
  2 Hyoid bone
  3 Cervical region
  4 Scapula
  5 Clavicle
  6 Sternum
  7 Thoracic region
  8 Humerus
  9 Radius
 10 Ulna
 11 Carpal bones
 12 Metacarpal bones
 13 Proximal phalanges
 14 Middle phalanges
 15 Distal phalanges
 16 Lumbar region
 17 Sacrum
 18 Coccygeal region
 19 Ilium
 20 Ischium
 21 Pubis
 22 Femur
 23 Patella
 24 Tibia
 25 Fibula
 26 Talus
 27 Calcaneus
 28 Metatarsal bones

Pelvic appendages The caudal or inferior appendages.

   Pelvic girdle Again, a partial girdle on each side, consisting of a "hip bone," the innominate. The two innominate bones articulate at the midline (forming the pubic symphysis), and each bone articulates with the sacrum (forming the sacroiliac joint on each side). The sacrum completes the girdle dorsally. Most anatomists refer to a single girdle in the pelvic region.
   Pelvic extremities The thigh, leg, and foot on each side.

   Note the differences between cat and human skeletons in the following:

   Clavicle
   Sternum
   Number of vertebrae in different regions
   Shape of innominate bone
   Number of digits (toes) on pelvic appendage

SOME BASIC BONE FEATURES

The various projections, lines, depressions, or surfaces on bones provide places for muscle or ligament attachment. It will be useful to become familiar with these if you will be studying the origins and insertions of muscles in detail. There are also openings for various purposes. Some general features are described below.

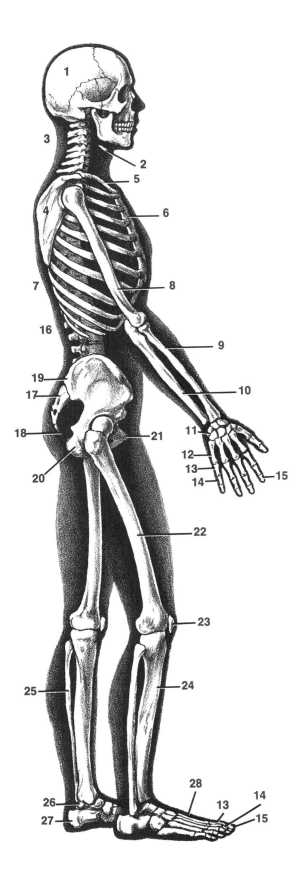

Figure 1-3
Human skeleton, lateral view

1 Skull
2 Hyoid bone
3 Cervical region
4 Scapula
5 Clavicle
6 Sternum
7 Thoracic region
8 Humerus
9 Radius
10 Ulna
11 Carpal bones
12 Metacarpal bones
13 Proximal phalanges
14 Middle phalanges
15 Distal phalanges
16 Lumbar region
17 Sacrum
18 Coccygeal region
19 Ilium
20 Ischium
21 Pubis
22 Femur
23 Patella
24 Tibia
25 Fibula
26 Talus
27 Calcaneus
28 Metatarsal bones

Aditus The entrance to a cavity.

Alveolus A deep pit or socket, such as one that holds a tooth.

Condyle A rounded or knuckle-like prominence, such as the occipital condyle; may be found at the articulation point of one bone with another.

Crest A narrow ridge of bone, as at the edge of the flared part of the hip bone.

Eminence A ridge or projection, or rounded prominence, like the rounded portions of the forehead bone.

Epicondyle A protuberance or prominence above a condyle, as above the condyles at the distal end of the humerus.

Facet A smooth, flattened articular surface, as in the gliding joints.

Fissure A narrow slit or cleft between bones, such as the orbital fissure.

Foramen An orifice for passage of blood vessels and/or nerves.

Fossa A depression or hollow, such as the orbital fossa for the eyeball.

Head A rounded projection beyond a constricted part or neck, as on the femur.

Line A ridge of bone less prominent than a crest; usually a site of muscle attachment.

Meatus, or **canal** A long, tube-like passage, such as the external ear canal.

Process *Any* marked bony prominence or projection.

Sinus, or **antrum** A cavity within a bone, such as the sinus in the frontal bone.

Spine A sharp, slender projection, such as the spinous process of a vertebra.

Sulcus A furrow or groove, such as the bicipital groove on the humerus for passage of a tendon.

Trochanter A large process for muscle attachment, such as the greater trochanter of the femur.

Trochlea A process shaped like a pulley, such as the one on the humerus that forms a joint with the ulna.

Tubercle A small rounded process, such as the tubercle of a rib.

Tuberosity A large rounded process, such as the ischial tuberosity.

DETAILS OF THE AXIAL DIVISION OF THE SKELETON

Using the human bones and the illustrations provided, identify the components of the different portions of the axial division of the skeleton. Some bones are paired (2), and some are single (1). Because many bones begin as two or more in embryonic development and then fuse, a bone listed as single may be present in two or more parts. The degree of fusion can vary; there may be fewer or more than is typical. It is not uncommon to see two frontal bones (the cat has two) or a single parietal bone where the bones have fused across the midline. Determine the major kinds of joints that are present.

Figure 1-4
Skull, front view

 1 Coronal suture
 2 Supraorbital foramen
 3 Optic canal (foramen)
 4 Superior orbital fissure
 5 Nasal bone
 6 Middle concha
 7 Perpendicular plate of ethmoid
    bone
 8 Inferior concha
 9 Vomer
10 Mandible
11 Frontal bone
12 Glabella
13 Parietal bone
14 Eminence
15 Supraorbital margin
16 Temporal bone
17 Lesser wing of sphenoid bone
18 Greater wing of sphenoid bone
19 Ethmoid bone
20 Lacrimal bone
21 Zygomatic arch
22 Infraorbital foramen
23 Zygomatic (malar) bone
24 Styloid process
25 Maxilla
26 Mental foramen

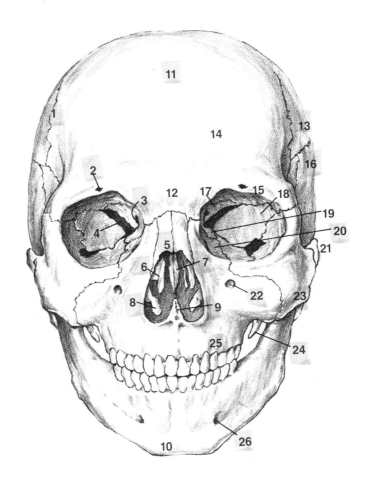

Skull
(Figs. 1-4, 1-5, 1-6, 1-7, 1-8, 1-9, 1-10, 1-11)

The skull typically contains 29 bones, including the hyoid.

Cranium

    Frontal—1
    Parietal—2
    Temporal—2
    Occipital—1
    Sphenoid—1
    Ethmoid—1

Ear ossicles (will not be seen)

    Malleus (hammer)—2
    Incus (anvil)—2
    Stapes (stirrup)—2

Hyoid—1

Face

    Maxillary—2
    Mandible—1
    Zygomatic, or **malar**—2
    Lacrimal—2
    Palatine—2
    Nasal—2
    Vomer—1
    Inferior concha, or **turbinate**—2

Note that some of the bones listed for the cranium could also be listed as facial bones.

Figure 1-5
Skull, side view

1 Lambdoidal suture
2 External occipital protuberance
3 Mastoid process
4 External auditory meatus
5 Styloid process
6 Zygomatic arch
7 Mental foramen
8 Eminence
9 Parietal bone
10 Squamous suture
11 Temporal bone
12 Mandible
13 Maxilla
14 Zygomatic bone
15 Lacrimal bone
16 Ethmoid bone
17 Greater wing of sphenoid bone
18 Frontal bone
19 Coronal suture
20 Occipital bone
21 Supraorbital margin
22 Nasal bone

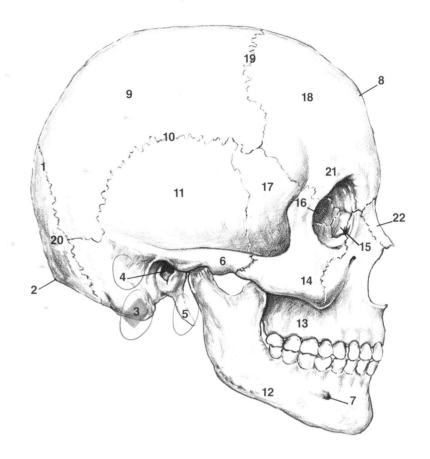

Special features of the skull

Sutures

  Sagittal Between the parietal bones.
  Coronal Between the frontal and parietal bones.
  Lambdoidal Between the occipital and parietal bones.
  Squamosal Between the squamosal portion of the temporal bone and other bones.
  Various others

Junctions that mark location of major fontanels in early life

  Bregma Junction of the frontal and the two parietal bones, at the anterior or frontal fontanel location.
  Lambda Junction of the occipital and the two parietal bones, at the posterior or occipital fontanel location.

Zygomatic arch The bony arch ventral to the ear, formed by portions of the temporal and zygomatic bones.

External auditory meatus The external ear canal. Note the opening.

Foramen magnum The large orifice in the occipital bone for passage of the spinal cord.

Occipital condyles The prominences that articulate with the first cervical vertebra.

Figure 1-6
Skull of newborn infant, side view

1 Parietal bone
2 Occipital (posterior) fontanel
3 Occipital bone
4 Mastoid (posterolateral) fontanel
5 Frontal (anterior) fontanel
6 Coronal suture
7 Sphenoid (anterolateral) fontanel
8 Frontal bone

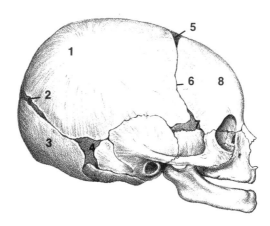

Mastoid process The prominence located dorsal and caudal to the opening of the external auditory meatus; it contains air cells, some of which open into the middle ear cavity.

Foramina other than foramen magnum Various passages for nerves and/or blood vessels.

Orbital fossae The hollows containing the eyeballs and associated structures.

Forehead That part of the frontal bone above the eyes and nose.

Nasal cavity Note the septum that divides the cavity. Because the nasal septum and the outer nasal skeleton are completed by cartilage, which is not present in the laboratory specimen, the "nose" will appear quite abbreviated.

Cranial cavity The space enclosed by the cranium that houses the brain.

Sinuses The cavities within the frontal, maxillary, ethmoid, and sphenoid bones. These are called the **paranasal** sinuses. (The bone is sponge-like within the lateral part of the ethmoid; rather than a single large sinus there are air cells, some of which open into the nasal cavity.)

Temporomandibular joint A combination hinge and gliding joint (ginglymoarthrodial). In the living body the joint cavity is divided by a complete intra-articular disk.

Details of the skull bones

Identify the special features of the various bones listed below. Use the illustrations provided.

Frontal

> **Supraorbital margin**
> **Supraorbital notch, or foramen**
> **Frontal sinuses** (within the bone)
> **Frontal eminence**(s) (The cat has a suture joint between two frontal bones.)

Parietal

> **Parietal eminence**
> **Temporal lines**
> **Foramina**

Figure 1-7
Base of skull

 1 Horizontal plate of palatine bone
 2 Temporal bone
 3 Foramen ovale
 4 Foramen spinosum
 5 Foramen lacerum
 6 Stylomastoid foramen
 7 Digastric groove
 8 Jugular fossa and foramen
 9 Occipital condyle
10 Occipital bone
11 Superior nuchal line
12 Inferior nuchal line
13 Incisive foramen
14 Palatine process of maxilla
15 Zygomatic arch
16 Medial pterygoid plate of sphenoid bone
17 Lateral pterygoid plate of sphenoid bone
18 Vomer
19 Styloid process
20 Mastoid process
21 Carotid canal
22 Foramen magnum
23 Median nuchal ine
24 External occipital protuberance

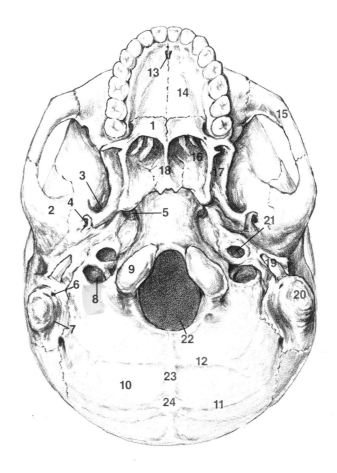

Temporal This consists of three main parts: a rather thin, squamous portion, a tympanic portion that forms much of the skeletal framework for the external auditory meatus (external ear canal), and a petromastoid portion usually described as two separate parts. The petrous portion is very hard and contains the middle and inner ears; the mastoid portion contains air cells that communicate with the tympanic cavity (middle ear cavity). The styloid process is a fourth part of the temporal bone. Identify the following:

Mastoid process
External auditory meatus The external ear canal.
Mastoid notch, or **digastric groove** Point of muscle attachment.
Zygomatic process Part of the zygomatic arch, or zygoma.
Mandibular fossa For articulation with the mandibular condyle.
Styloid process Usually broken off; gives attachment to the hyoid bone by ligament and also to muscle.
Carotid canal Gives passage to the internal carotid artery.
Jugular foramen Gives passage to the internal jugular vein and some of the cranial nerves.
Stylomastoid foramen Gives passage to a cranial nerve (the facial) that supplies the muscles of facial expression. (This nerve is affected in "Bell's palsy," a unilateral paralysis of these muscles of expression.)
Internal auditory meatus Gives passage to the nerves and blood vessels that supply the ear.

Figure 1-8
Floor of cranial cavity

1 Optic foramen
2 Foramen rotundum
3 Foramen ovale
4 Foramen lacerum
5 Jugular foramen
6 Internal auditory meatus
7 Foramen magnum
8 Crista galli
9 Cribriform plate of ethmoid
  bone
10 Frontal bone
11 Anterior cranial fossa
12 Lesser wing of sphenoid bone
13 Greater wing of sphenoid bone
14 Temporal bone
15 Middle cranial fossa
16 Sella turcica
17 Petrous portion of temporal
  bone
18 Hypoglossal canal
19 Posterior cranial fossa
20 Parietal bone
21 Occipital bone
22 Foramen spinosum
23 Posterior clinoid process
24 Anterior clinoid process

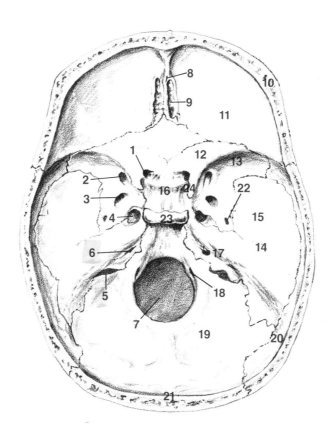

Occipital

Foramen magnum For passage of the spinal cord, blood vessels, and the spinal root of the eleventh cranial nerve.

Occipital condyles Articulate with the first cervical vertebra (the atlas) to form the atlantoöccipital joints.

Nuchal lines (superior, inferior, and median) Give attachment to muscles.

External occipital protuberance The raised area between the superior nuchal lines; it can be palpated at the midline.

Foramina other than magnum

Sphenoid

Body The central portion of the bone.

Greater and lesser wings Lateral projections.

Sella turcica The depression lodging the hypophysis.

Sphenoid sinuses Located within the body of the bone.

Pterygoid processes Note the plates of each process, and the fossa between the plates.

Optic foramen For passage of the optic nerve, the nerve of vision.

Other foramina

Ethmoid

Cribriform plate A horizontal plate with perforations for passage of olfactory nerve fibers. The olfactory nerve is the nerve for smell.

Perpendicular plate The upper part of the nasal septum.

Figure 1-9
Skull, sagittal section

1 Frontal bone
2 Parietal bone
3 Occipital bone
4 Internal auditory meatus
5 Hypoglossal canal
6 Styloid process
7 Pterygoid process
8 Vomer
9 Palatine bone
10 Palatine process of maxilla
11 Incisive canal
12 Temporal bone
13 Sella turcica
14 Sphenoid sinus
15 Crista galli
16 Frontal sinus
17 Nasal bone
18 Perpendicular plate of ethmoid bone
19 Lambdoidal suture
20 Squamous suture
21 Maxilla
22 Coronal suture

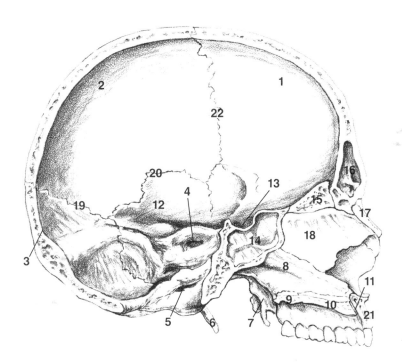

Lateral mass Located in the medial wall of the orbital fossa; contains air cells (ethmoid sinus).

Superior and middle conchae, or **turbinates** Located in the lateral wall of the nasal cavity.

Maxillary (upper jaw)

Alveolar portion Contains cavities for teeth.

Maxillary sinus Located within the bone.

Palatine process Part of the hard palate.

Infraorbital foramen Gives passage to the infraorbital nerve and blood vessels.

Lacrimal groove

Zygomatic, or **malar** (the "cheekbone")

Lacrimal Note the opening to the lacrimal canal, which contains the tear duct in the living body. The canal terminates in the nasal cavity.

Palatine The horizontal plates of the two palatine bones form part of the hard palate.

Nasal The two nasal bones form the bridge of the nose.

Vomer Part of the lower portion of the nasal septum.

Inferior concha, or **turbinate** Small bone below the middle concha of the ethmoid. Because these conchae are very fragile, due to the spongy character of the bone, they are often missing from the skulls in the laboratory.

Mandible (lower jaw)

Symphysis The site at which the halves of the mandible fuse. (The cat has a suture joint between the halves.)

Figure 1-10
Mandible

 1 Neck of condyle
 2 Condyloid process
 3 Ramus
 4 Angle
 5 Body
 6 Mental foramen
 7 Alveolar border
 8 Mandibular foramen
 9 Notch
10 Coronoid process
11 Mental symphysis
12 Lingula
13 Mylohyoid groove

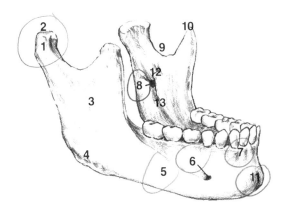

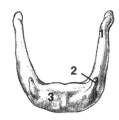

Figure 1-11
Hyoid bone, front view

1 Greater cornu
2 Lesser cornu
3 Body (basihyal)

Body The main portion of the bone.
Ramus
Angle
Condyle, or **condyloid process** Articulates with the temporal bone.
Neck of condyle
Coronoid process Note the mandibular notch between this process and the condyle.
Alveolar portion Contains cavities for teeth.
Foramina

Hyoid The hyoid is not present on a separate skull, but it should be on an articulated skeleton. It can be observed deep in the "horseshoe curve" of the mandible. It forms no joints with other bones; it is attached only by a ligament to the styloid process of the temporal bone, and is held in place by muscle attachments.

Vertebral Column
(Figs. 1-12, 1-13, 1-14, 1-15, 1-16, 1-17, 1-18, 1-19, 1-20)

The vertebral column in the adult human typically contains 26 bones.

Number of vertebrae

The number of vertebrae present, in human and cat, are listed below by region.

	Number in human	Number in cat
Cervical	7	7
Thoracic	12	13
Lumbar	5	7
Sacral	5 (fused to form one sacrum)	3 (fused to form one sacrum)
Coccygeal, or caudal	2–5 (fused to form one coccyx)	many

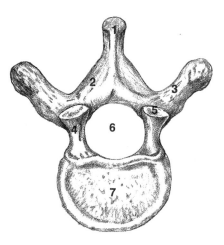

Figure 1-12
A typical vertebra

1 Spinous process
2 Lamina
3 Transverse process
4 Pedicle
5 Articular process
6 Vertebral foramen
7 Body

Features of a typical vertebra

Body, or **centrum**

Neural arch Formed by two pedicles and two fused laminae (see below).

Dorsal spine, or **spinous process**

Lateral spines, or **transverse processes**

Vertebral foramen When vertebrae are fitted together, the foramina participate in the formation of a canal, the **neural canal,** that houses the spinal cord.

Laminae Together these support the spinous process and form the dorsal wall of the neural arch.

Pedicles The "roots" of the neural arch.

Superior articular processes and facets

Inferior articular processes and facets

Vertebral notches These are distinguished as "superior" and "inferior" and are located in the pedicles. (When two vertebrae are fitted together, two opposing notches form an intervertebral foramen for passage of a spinal nerve and intervertebral blood vessels.)

Articulations between vertebrae

Each vertebra participates in forming joints with the one(s) adjacent to it, so that a "chain" of bones is formed. On the articulated skeleton, observe the gliding joints between inferior and superior articular processes on the neural arches of adjacent vertebrae. Note the felt pads that simulate the intervertebral disks. These disks are made of fibrocartilage (a compressible tissue) and act as shock absorbers. They form symphysis joints with the bodies of the vertebrae and are the chief means of connection between vertebral bodies. (There are also various ligamentous connections between vertebrae and other ligaments running the entire length of the vertebral column, and you should read the descriptions of these in your textbook.)

Because of variations in the intervertebral disks and vertebral bodies, four spinal curvatures are formed: **cervical, thoracic, lumbar,** and **sacral** (Fig. 1-13). The intervertebral disks vary in thickness in different regions of the vertebral column, with those in the lumbar region being the thickest and those in the thoracic region being the thinnest. In the lumbar and cervical regions, the disks are wedge-shaped, with the thick side being ventral and the thin side dorsal, so they are of primary importance in formation of the cervical and lumbar curvatures. These are the regions of greatest mobility of the vertebral column, and the thick lumbar disks are the most likely to rupture, especially the last one between the fifth lumbar vertebra and the sacrum, it being much thicker ventrally than it is dorsally. In the thoracic and sacral regions, the variations in the shapes of vertebral

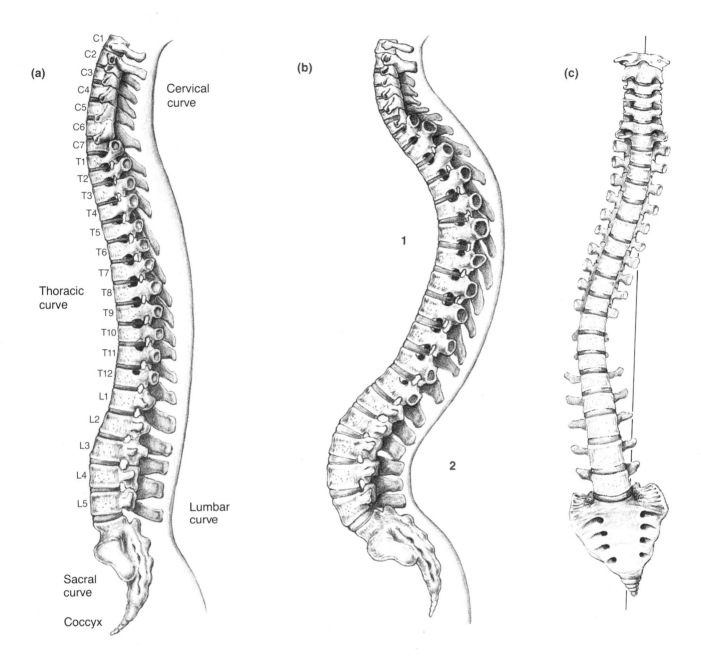

(a)

C1
C2
C3
C4
C5
C6
C7
T1
T2
T3
T4
T5
T6
T7
T8
T9
T10
T11
T12
L1
L2
L3
L4
L5

Cervical curve

Thoracic curve

Lumbar curve

Sacral curve

Coccyx

(b)

1

2

(c)

Figure 1-13
Curvatures of the vertebral column
(a) Normal curvatures
(b) Kyphosis 1; lordosis 2
(c) Scoliosis

bodies are primarily responsible for the curvatures. There are no disks between sacral vertebrae, since the vertebrae are fused, but there is usually a small one between the sacrum and the coccyx.

The normal curvatures of the vertebral column are very important for the flexibility of an upright posture; a straight column would not be so flexible. However, the curvatures are sometimes abnormally pronounced, or veer from the midline, and some of the skeletons that you will see in the laboratory may show one or more of these deformities. Common deformities (Fig. 1-13) are: **kyphosis,** an exaggerated thoracic curvature; **lordosis,** an exaggerated lumbar curvature; and **scoliosis,** a curvature that veers to either side of the midline of the body (a lateral curvature).

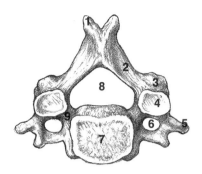

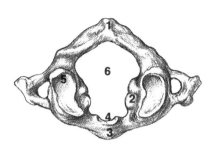

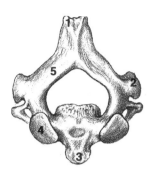

Figure 1-14
A typical cervical vertebra (fourth), cranial aspect

1 Spinous process
2 Lamina
3 Inferior articular process
4 Superior articular process
5 Transverse process
6 Transverse foramen
7 Body
8 Vertebral foramen
9 Pedicle

Figure 1-15
First cervical vertebra, cranial aspect

1 Posterior tubercle
2 Tubercle for transverse ligament
3 Anterior tubercle
4 Facet for odontoid process of axis
5 Superior articular facet
6 Vertebral foramen

Figure 1-16
Second cervical vertebra, cranial aspect

1 Spinous process
2 Inferior articular process
3 Odontoid process (dens)
4 Superior articular facet
5 Lamina

Special features of regional vertebrae

Using the illustrations provided, identify the features of a typical vertebra from each region of the vertebral column.

CERVICAL REGION Note the short, bifid spinous process of a typical cervical vertebra. Because the seventh cervical has a spine resembling those of the thoracic vertebrae and can be palpated projecting out at the base of the neck, it is called the **vertebra prominens.** Note the **transverse foramen** on each side. These foramina, which are distinctive to the cervical vertebrae, give passage to the vertebral artery and vein.

The first two cervical vertebrae are different from the others. The first one is called the **atlas;** and the second, the **axis,** or **epistropheus.** These two have a special articulation between them that is not present in the others. The first also has special articular surfaces for the occipital condyles. Find these articular surfaces, and note also that the centrum and spinous process are greatly underdeveloped. Find the articular surface for the **dens,** or **odontoid process,** of the axis. On the second cervical find the dens, as well as other features typical of a cervical vertebra. The articulations between the atlas and the occipital bone are called the **atlantoöccipital joints.** These are condyloid joints. The articulation between the atlas and the dens of the axis is called the **atlantoaxial,** or **atlantoepistropheal, joint.** This is a pivot, trochoid, or rotary joint.

THORACIC REGION Note the long spinous process of a typical thoracic vertebra. Note articular facets for ribs on the body and the transverse processes. These articular surfaces for ribs distinguish a vertebra of the thoracic region. The long, sharp spine is also a diagnostic feature of all but the last one or two vertebrae, which have a spine resembling those of the lumbar region.

Figure 1-17
Thoracic vertebra

(a) Cranial aspect
(b) Lateral aspect
1 Spinous process
2 Lamina
3 Facet for tubercle of rib
4 Superior articular process
5 Superior demifacet for head
   of rib
6 Body
7 Pedicle
8 Transverse process
9 Inferior articular process
10 Superior notch
11 Inferior notch
12 Inferior demifacet for head
    of rib

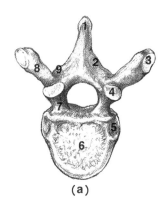

(a)

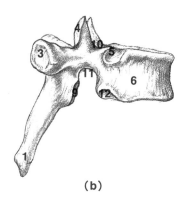

(b)

Typically, the head of a rib articulates with demifacets (half facets) on the bodies of two adjacent vertebrae, and the tubercle of the rib articulates with the facet on the transverse process of the caudally adjacent vertebra. With the first rib and the last two, the head articulates with the body of only one vertebra and the tubercle with the transverse process of that same vertebra.

LUMBAR REGION Note the broad, blunt spinous process and the comparatively slender transverse processes of a lumbar vertebra. The bodies of the lumbar vertebrae are larger than those of the thoracic vertebrae. (The bodies of the vertebrae are increasingly larger along the column in the caudal direction up to the sacral region, where the vertebrae fuse.)

SACRAL REGION Five vertebrae (three in the cat) have fused to form the **sacrum,** which provides the dorsal wall of the pelvic cavity. Find the features listed:

Body Note the fusion lines on the ventral surface.

Promontory Ventral projection of the first sacral vertebra.

Alae, or **lateral masses**

Sacral foramina: anterior (ventral) and **posterior** (dorsal). For passage of ventral and dorsal rami of sacral spinal nerves.

Figure 1-18
Lumbar vertebra
(a) Cranial aspect
(b) Lateral aspect

1 Spinous process
2 Lamina
3 Inferior articular process
4 Superior articular process
5 Transverse process
6 Pedicle
7 Body

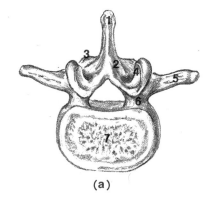

(a)

(b)

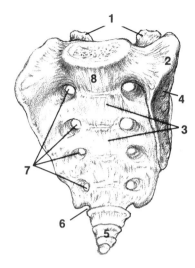

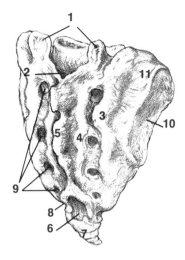

Figure 1-19
Sacrum and coccyx, ventrolateral view

1 Superior articular processes
2 Lateral mass
3 Body
4 Articular surface
5 Coccyx
6 Apex of sacrum
7 Anterior sacral foramina
8 Promontory

Figure 1-20
Sacrum and coccyx, dorsolateral view

1 Superior articular processes
2 Neural canal
3 Lateral sacral crest
4 Intermediate sacral crest
5 Median sacral crest
6 Hiatus
7 Coccyx
8 Apex of sacrum
9 Posterior sacral foramina
10 Articular surface
11 Lateral mass

Superior articular processes and facets For articulation with the most caudal lumbar vertebra.

Articular surfaces for the ilium (for sacroiliac joints)

Lateral crests These are lateral to the posterior foramina, and represent transverse processes.

Median crest The crest at the median line that represents spinous processes.

Apex The caudal limit of the sacrum.

Sacral hiatus Located on the dorsal side.

Sacral portion of the neural canal

Note the differences in the sacrum of male and female. The sacrum in the male is narrow and more curved; in the female it is broad and less curved.

COCCYGEAL REGION The vertebrae are rudimentary and are usually fused to form the coccyx, although the first one is often separate.

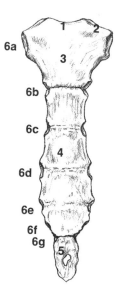

Figure 1-21
Sternum, ventral surface

1 Jugular notch
2 Clavicular notch
3 Manubrium
4 Body
5 Xiphoid process
6 Costal notches
  a First costal notch
  b Second costal notch
  c Third costal notch
  d Fourth costal notch
  e Fifth costal notch
  f Sixth costal notch
  g Seventh costal notch

Thorax

(Figs. 1-21, 1-22)

The human thorax contains 25 bones (in addition to 12 thoracic vertebrae).

Note that the thorax is somewhat cone-shaped, being narrower at the cranial end, the apex, than at the caudal end, the base. In addition to the thoracic vertebrae, the bones of the thorax are the **sternum,** in the ventral midline region, and the **ribs** (12 pairs, or 24). The bones form a bony cage that protects the thoracic viscera.

Sternum

Identify the parts:

Manubrium The most cranial part.
Body, or **gladiolus** The middle part.
Xiphoid process The most caudal part. This is cartilaginous in the young
  person, but becomes calcified in the older person.

Observe the following features:

Suprasternal, or **jugular notch** This indentation can be palpated at midline, at the cranial extremity of the sternum.
Clavicular notch The site at which the clavicle articulates.
Costal notches The sites at which the ribs articulate.
Sternal angle A slight ridge that can be palpated at the junction of the
  manubrium and the body; this angle marks the level at which the
  costal cartilage of the second rib joins the sternum.

Note the articulations of the sternum:

Sternoclavicular A gliding joint.
Sternocostal All are gliding joints, except the first, which is
  a synchondrosis (a cartilaginous joint).

Ribs

Identify the parts:

Body, or **shaft** The long part of the rib.
Head (at dorsal extremity)
Neck (at dorsal extremity)
Tubercle (at dorsal extremity)
Costal cartilage (at ventral extremity)

Observe the features:

Angle (near dorsal extremity)
Articular facets on the head and tubercle

Note the articulations of the ribs:

Sternocostal Noted above.
Costovertebral All are gliding joints.

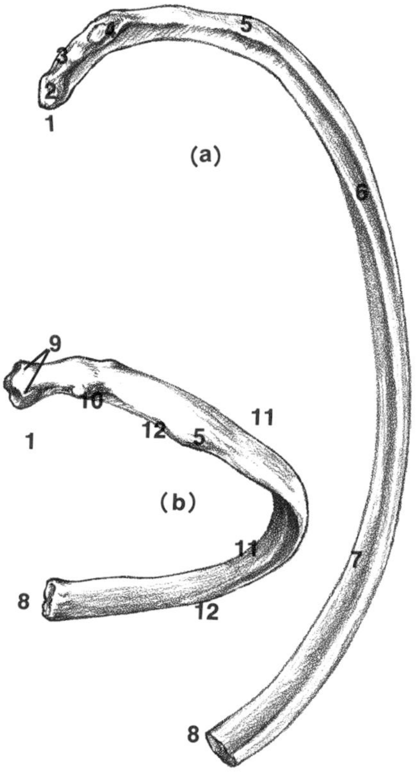

Figure 1-22
A typical rib (right)
   (a) Inferior surface
   (b) Dorsal view
The costal cartilage is not shown.

 1 Dorsal (vertebral) extremity
 2 Head
 3 Neck
 4 Tubercle
 5 Angle
 6 Costal groove
 7 Body
 8 Ventral extremity
 9 Articular facets for bodies of vertebrae
10 Articular facet for a transverse process
11 Cranial border
12 Caudal border

Ribs 1 through 7 are called **true,** or **sternal** ribs because they articulate directly with the sternum via their costal cartilages. Note that the first rib is much shorter, broader, and more curved than the others. Ribs 8, 9 and 10 are called **false,** or **asternal** ribs because they do not articulate directly with the sternum. The costal cartilage of each articulates with the costal cartilage of the rib just cranial to it. Ribs 11 and 12 are also false, or asternal ribs, but they are called "floating ribs" because their ventral ends are free.

Note the curved downward slope of the ribs from the vertebrae to the ventral side of the body, and the upward (cranial) swing of the costal cartilage. A transverse section through the thorax would cut across more than one rib.

(A thirteenth rib, known as a "gorilla rib," is occasionally present in the human. Its name is derived from the fact that gorillas usually have 13 pairs of ribs. Cats, too, typically have 13 pairs of ribs.)

DETAILS OF THE APPENDICULAR DIVISION OF THE SKELETON

Proceed with this study as you did for the axial division of the skeleton.

Pectoral Appendages
(Figs. 1-23, 1-24, 1-25, 1-26, 1-27, 1-28, 1-29, 1-30)

In the adult human, the pectoral appendages contain 64 bones.

Pectoral girdles These attach the bones of the pectoral extremities to the axial division of the skeleton.

   Scapula (shoulder blade)—1 on each side.
   Clavicle (collar bone)—1 on each side.

Pectoral extremities
   Arms
      Humerus—1 in each arm.
   Forearms
      Radius (on thumb side)—1 in each forearm.
      Ulna (on little-finger side)—1 in each forearm.
   Hands
      Carpals (wrist bones)—8 in each hand.
      Metacarpals—5 in each hand.
      Phalanges (bones of the digits)—2 in each thumb, 3 in each finger.

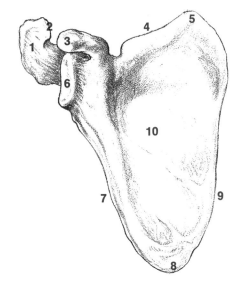

Figure 1-24
Scapula (right), ventral aspect

1 Acromion
2 Facet for clavicle
3 Coracoid process
4 Superior border
5 Superior angle
6 Glenoid cavity
7 Lateral border
8 Inferior angle
9 Medial border
10 Subscapular fossa

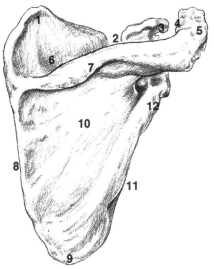

Figure 1-23
Scapula (right), dorsal aspect

1 Superior angle
2 Suprascapular notch
3 Coracoid process
4 Facet for clavicle
5 Acromion
6 Supraspinous fossa
7 Spine
8 Medial border
9 Inferior angle
10 Infraspinous fossa
11 Lateral border
12 Infraglenoid tubercle

Figure 1-25
Clavicle (right), cranial aspect

1 Acromial extremity
2 Conoid, or coracoid tubercle
3 Deltoid tubercle
4 Sternal extremity

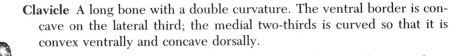

Details of bones of the pectoral appendage

Scapula A large, flat, triangular bone.

Borders
    Superior, or **cranial**
    Lateral, or **axillary**
    Medial, or **vertebral**
Angles
    Superior, or **cranial**
    Inferior, or **caudal**
    Lateral, or **acromial**
Fossae
    Supraspinous Dorsal; craniad of the spine.
    Infraspinous Dorsal; caudad of the spine.
    Subscapular Ventral.
    Glenoid Faces laterally; articulates with the head of the humerus.
Other features
    Acromion
    Spine
    Coracoid process
    Scapular notch Located in the superior border.
    Infraglenoid tubercle

Clavicle A long bone with a double curvature. The ventral border is concave on the lateral third; the medial two-thirds is curved so that it is convex ventrally and concave dorsally.

Sternal end Triangular, with the apex directed downward; it articulates with the manubrium of the sternum.
Acromial end This is more flat than the sternal end; it articulates with the acromion of the scapula.
Coracoid, or **conoid tubercle** Located near the acromial end, dorsal and somewhat caudal; it is attached by a ligament to the coracoid process of the scapula.
Deltoid area A rough area at the ventral border of the acromial end; part of the deltoid muscle arises here.

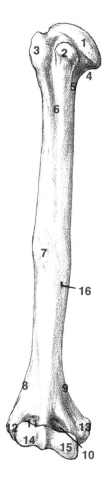

Figure 1-26
Humerus (right), ventral surface

1 Head
2 Lesser tubercle
3 Greater tubercle
4 Anatomical neck
5 Surgical neck
6 Intertubercular groove
7 Deltoid tuberosity
8 Lateral supracondylar ridge
9 Medial supracondylar ridge
10 Coronoid fossa
11 Radial fossa
12 Lateral epicondyle
13 Medial epicondyle
14 Capitulum (lateral condyle)
15 Trochlea (medial condyle)
16 Nutrient foramen

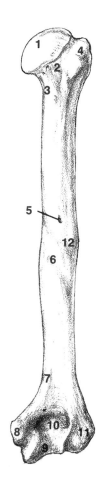

Figure 1-27
Humerus (right), dorsal surface

1 Head
2 Anatomical neck
3 Surgical neck
4 Greater tubercle
5 Nutrient foramen
6 Groove for radial nerve
7 Medial supracondylar ridge
8 Medial epicondyle
9 Trochlea
10 Olecranon fossa
11 Lateral epicondyle
12 Deltoid tuberosity

Humerus

Head Located on the medial side.
Greater tubercle, or **tuberosity**
Lesser tubercle, or **tuberosity**
Anatomical neck A slight constriction between the head and the tubercles and shaft.
Surgical neck A constriction of the shaft below the head and tubercles.
Intertubercular, or **bicipital groove** For the passage of a tendon of the biceps brachii muscle.
Deltoid tuberosity An elevation on the ventrolateral surface. This is the site of insertion for the deltoid muscle.
Spiral, or **musculospiral,** or **radial groove** A shallow, oblique depression crossed by the radial nerve.
Medial and lateral epicondyles
Medial and lateral supracondylar ridges
Lateral condyle, or **capitulum** Articulates with the head of the radius.
Medial condyle, or **trochlea** Articulates with the semilunar notch of the ulna.
Olecranon fossa Located dorsally at the distal end; it receives an ulnar process, the olecranon, when the forearm is extended.
Coronoid fossa Located ventrally at the distal end, superior to the trochlea; it receives the coronoid process of the ulna when the forearm is flexed.

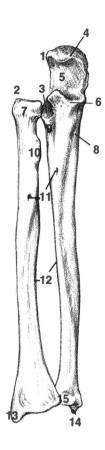

Figure 1-28
Ulna and radius (right), ventral surfaces

1 Ulna
2 Radius
3 Radial notch of ulna
4 Olecranon process
5 Semilunar notch
6 Coronoid process
7 Head of radius
8 Ulnar tuberosity
9 Supinator crest
10 Radial tuberosity
11 Nutrient foramina
12 Interosseous margins
13 Styloid process of radius
14 Styloid process of ulna
15 Head of ulna

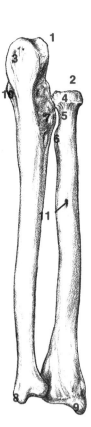

Figure 1-29
Ulna and radius (right), dorsal surfaces

1 Ulna
2 Radius
3 Olecranon
4 Articular circumference of head of radius
5 Neck of radius
6 Radial tuberosity
7 Supinator crest
8 Styloid process of ulna
9 Styloid process of radius
10 Coronoid process
11 Nutrient foramen

Radial fossa Located ventrally at the distal end, superior to the capitulum; it receives the ventral portion of the head of the radius when the forearm is flexed.

Always note, in the bones, the nutrient foramina that blood vessels pass through.

Ulna The medial bone of the forearm; it has a triangular shape in cross section.

Olecranon process (elbow bone) Located dorsally at the proximal end.
Coronoid process Located ventrally at the proximal end.
Semilunar, or **trochlear notch** Articulates with the humerus.
Radial notch On the lateral side of the proximal end; it articulates with the radius.
Head Located at the distal end; it has an articular surface for the radius.
Styloid process Located at the distal end.
Interosseous margin, or **crest** Located on the lateral side.

Radius The lateral bone of the forearm.

Head Located at the proximal end. The shallow depression articulates with the lateral condyle, or capitulum, of the humerus; the circumference of the head articulates with the radial notch of the ulna.
Radial, or **bicipital tuberosity** A protuberance for insertion of the biceps brachii muscle.
Styloid process Located at the distal end, on the lateral side.
Ulnar notch Located at the distal end, on the medial side; it articulates with the head of the ulna.
Interosseous margin, or **crest** Located on the medial side.

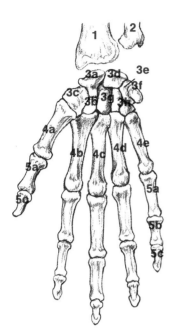

Figure 1-30
Bones of hand (right), palmar surface

1 Radius
2 Ulna
3 Carpals
  a Navicular (scaphoid)
  b Trapezoideum (lesser multangular)
  c Trapezium (greater multangular)
  d Lunate (semilunar)
  e Triquetrum (triangular)
  f Pisiform
  g Capitate
  h Hamate (unciform)
4 Metacarpals
  a First metacarpal
  b Second metacarpal
  c Third metacarpal
  d Fourth metacarpal
  e Fifth metacarpal
5 Phalanges
  a Proximal phalanx
  b Middle phalanx
  c Distal phalanx

Observe the distal end of the radius, noting the smooth flattened ventral surface and the ridged and grooved dorsal surface, and the articular surfaces for carpal bones.

Carpals These are listed from thumb side to little finger side for each row.

  Proximal row
    Scaphoid, or **navicular**
    Lunate
    Triangular, or **triquetrum**
    Pisiform Small, rounded elevation on the little finger side.
  Distal row.
    Trapezium, or **greater multangular**
    Trapezoideum, or **lesser multangular**
    Capitate
    Hamate

Metacarpals These are not named, but only numbered as follows, beginning with the thumb side: 1st, 2nd, 3rd, 4th, 5th.

Phalanges (singular: **phalanx**) Two in each thumb: proximal, distal. Three in each finger: proximal, middle, distal.

Articulations of the pectoral appendage

Sternoclavicular joint The gliding joint between the sternum and the clavicle.

Acromioclavicular joint The gliding joint between the acromion of the scapula and the clavicle.

Shoulder joint (glenohumeral or scapulohumeral) A ball-and-socket joint between the glenoid fossa of the scapula and the head of the humerus.

Elbow joint Between the medial and lateral condyles of the humerus, and the proximal ends of the ulna and radius (at the semilunar notch of the ulna and the head of the radius). Functionally, this is a hinge joint, but anatomically the radiohumeral component is a gliding joint.

Radioulnar joints

  Proximal That between the radial notch of the ulna, and the circumference of the head of the radius. It is a pivot, or trochoid, or rotary joint.
  Middle A fibrous interosseous membrane connects the shafts of the radius and ulna. This joint may be classified as a syndesmosis.
  Distal Between the ulnar notch of the radius and the head of the ulna; a pivot, or trochoid, or rotary joint.

Wrist joint Between the proximal carpals (except for the pisiform), and the distal end of the radius and a cartilage disk that extends from the radius over the distal end of the ulna. This is often classified as a condyloid joint, but some anatomists consider it ellipsoid because of the shallowness of the convex and concave articular surfaces.

Intercarpal joints Those between the carpals; all are gliding joints.

Carpometacarpal joints Between the carpals and the metacarpals. All are gliding joints except the one between the trapezium (greater multangular) and the first metacarpal, which is a saddle joint.

Intermetacarpal joints Between the metacarpals; these are gliding joints.

Metacarpophalangeal joints Between the metacarpals and the proximal phalanges; generally classified as condyloid joints, but sometimes as ball-and-socket.

Interphalangeal joints Between the phalanges; all are hinge joints.

Pelvic Appendages

(Figs. 1-31, 1-32, 1-33, 1-34, 1-35, 1-36, 1-37, 1-38, 1-39, 1-40, 1-41)

In the adult human, the pelvic appendages contain 62 bones.

Pelvic girdle Attaches the bones of the pelvic extremities to the axial division of the skeleton.

  Innominate bones—1 on each side. Each innominate bone is composed of three parts that are separately developed, but fused.
    Ilium The cranial, flared part.
    Ischium The caudal, dorsal part.
    Pubis The ventral part.

Pelvic extremities

  Thighs
    Femur—1 in each thigh.
  Legs
    Tibia (the large bone on the medial side)—1 in each leg.
    Fibula (the small lateral bone)—1 in each leg.
  Feet
    Tarsals—7 in each foot.
    Metatarsals—5 in each foot.
    Phalanges (bones of the digits)—2 in each large toe, and 3 in each of the other toes.
  Kneecap
    Patella—1 in each extremity.

Details of bones of the pelvic appendage

Innominate bones
  Ilium
    Iliac crest
    Iliac spines Distinguished as the anterior superior, anterior inferior, posterior superior, and posterior inferior.
    Iliac fossa Located on the inner surface.
    Gluteal lines Distinguished as the posterior, anterior or middle, and inferior; all are located on the outer surface.
    Greater sciatic notch
    Articular surface for the sacrum
  Ischium
    Ischial spine
    Ischial tuberosity
    Lesser sciatic notch
    Superior and inferior rami

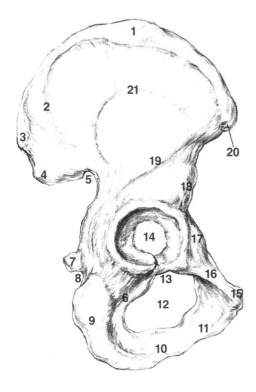

Figure 1-31
Innominate bone (right), external aspect

 1 Crest of ilium
 2 Posterior gluteal line
 3 Posterior superior iliac spine
 4 Posterior inferior iliac spine
 5 Greater sciatic notch
 6 Superior ramus of ischium
 7 Ischial spine
 8 Lesser sciatic notch
 9 Ischial tuberosity
10 Inferior ramus of ischium
11 Inferior ramus of pubis
12 Obturator foramen
13 Acetabular notch
14 Acetabulum
15 Pubic tubercle
16 Superior ramus of pubis
17 Iliopectineal eminence
18 Anterior inferior iliac spine
19 Inferior gluteal line
20 Anterior superior iliac spine
21 Anterior gluteal line

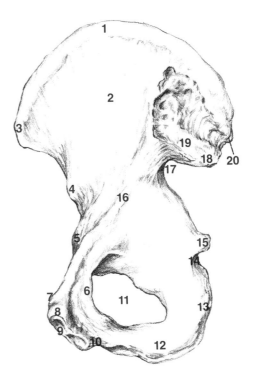

Figure 1-32
Innominate bone (right), internal aspect

 1 Crest of ilium
 2 Iliac fossa
 3 Anterior superior iliac spine
 4 Anterior inferior iliac spine
 5 Iliopectineal eminence
 6 Superior ramus of pubis
 7 Pubic tubercle
 8 Pubic crest
 9 Pubic symphysis (articular surface)
10 Inferior ramus of pubis
11 Obturator foramen
12 Inferior ramus of ischium
13 Ischial tuberosity
14 Lesser sciatic notch
15 Ischial spine
16 Arcuate line
17 Greater sciatic notch
18 Posterior inferior iliac spine
19 Articular surface for sacrum
20 Posterior superior iliac spine

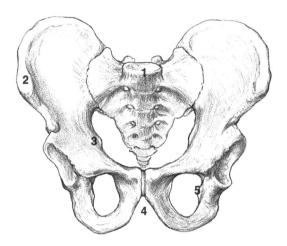

Figure 1-33
Male pelvis

1 Sacral promontory
2 Anterior superior iliac spine
3 Arcuate line
4 Pubic arch
5 Ischial spine

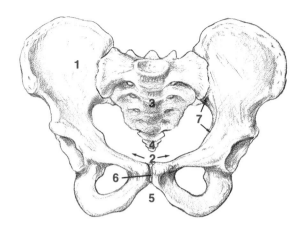

Figure 1-34
Female pelvis

1 Greater (false) pelvis
2 Lesser (true) pelvis
3 Sacrum
4 Coccyx
5 Pubic arch
6 Pubic symphysis
7 Brim of lesser pelvis

Pubis
 Superior and inferior rami
 Pubic arch The arch formed by the articulated pubic bones.
 Pubic symphysis The joint between the pubic bones.
Other features
 Acetabulum Note the lunate articular surface for articulation with the head of the femur.
 Obturator foramen

On an articulated pelvis, note the following:

Pelvic brim, or **inlet** The inlet to the true pelvis.
Pelvic outlet Caudal to the brim or inlet.
True (lesser) pelvis Between inlet and outlet.
False (greater) pelvis Cranial to the brim or inlet of the true pelvis.

Note also the differences between male and female in their articulated pelves. In most males the angle formed by the pubic arch is less than a right angle; in the female it is a right angle, or greater, so that the ischial tuberosities are turned outward more. In the female the ilium is flared more laterally, and the pelvic inlet is proportionately broader from side to side (left to right) than that of the male, so that it is more circular. The female sacrum is shorter, broader, and less curved than that of the male.

Femur

 Head Located on the medial side of the proximal end.
 Fovea capitis A depression in the head for ligament attachment. The ligament attaches the head of the femur to the acetabular fossa.

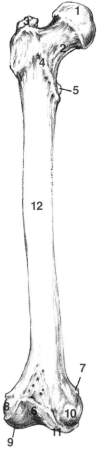

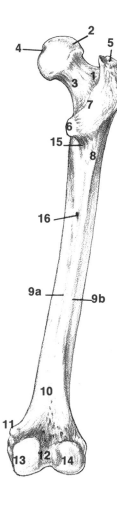

Figure 1-35
Femur (right), ventral surface

1 Head
2 Neck
3 Greater trochanter
4 Intertrochanteric line
5 Lesser trochanter
6 Patellar surface
7 Adductor tubercle
8 Lateral epicondyle
9 Lateral condyle
10 Medial epicondyle
11 Medial condyle
12 Shaft

Figure 1-36
Femur (right), dorsal surface

1 Trochanteric fossa
2 Head
3 Neck
4 Fovea capitis
5 Greater trochanter
6 Lesser trochanter
7 Intertrochanteric crest
8 Gluteal tuberosity
9 Linea aspera
   a Medial lip
   b Lateral lip
10 Popliteal surface
11 Adductor tubercle
12 Intercondylar fossa (notch)
13 Medial condyle
14 Lateral condyle
15 Pectineal line
16 Nutrient foramen

Neck Constriction between the head and the trochanteric prominences. This is the place most likely to be fractured in a fracture of the "hip."
Greater trochanter The prominence on the lateral side.
Lesser trochanter Located on the dorsomedial side.
Trochanteric fossa Located on the dorsal side, medial to the greater trochanter.
Intertrochanteric line On the ventral surface.
Intertrochanteric crest On the dorsal surface.
Linea aspera On the dorsal surface.
Medial and lateral condyles These articulate with the tibia.
Medial and lateral epicondyles
Adductor tubercle On the medial side at the distal end.
Intercondylar notch, or **fossa** The depression between the condyles.

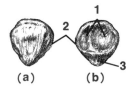

Figure 1-37
Patella (right)
   (a) Ventral surface
   (b) Dorsal surface

1 Articular surfaces for femur
2 Medial border
3 Point of attachment of patellar ligament

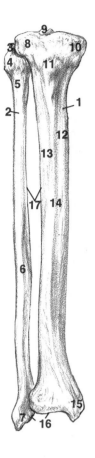

Figure 1-38
Tibia and fibula (right), ventral surfaces

1 Tibia
2 Fibula
3 Styloid process of fibula
4 Head of fibula
5 Neck of fibula
6 Anterior crest of fibula
7 Lateral malleolus
8 Lateral condyle of tibia
9 Intercondylar eminence
10 Medial condyle of tibia
11 Tibial tuberosity
12 Medial surface of tibia
13 Lateral surface of tibia
14 Anterior crest of tibia
15 Medial malleolus
16 Articular surfaces for talus
17 Interosseous borders

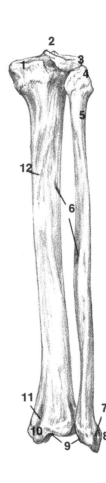

Figure 1-39
Tibia and fibula (right), dorsal surfaces

1 Medial condyle of tibia
2 Intercondylar eminence
3 Lateral condyle of tibia
4 Head of fibula
5 Neck of fibula
6 Nutrient foramen
7 Groove for tendons of peroneus longus and peroneus brevis
8 Lateral malleolus
9 Articular surfaces for talus
10 Medial malleolus
11 Groove for tendons of tibialis posterior and flexor digitorum longus
12 Soleal (popliteal) line

Trochlea (patellar surface)
Popliteal surface The dorsal surface, at the distal end.

Patella The kneecap; it is a sesamoid bone.

Tibia (shin bone)

Medial and lateral condyles These articulate with the condyles of the femur.

Intercondylar eminence

Tibial tuberosity A rounded process on the ventral surface.

Anterior, or tibial crest Located on the ventral surface.

Popliteal, or soleal line On the dorsal surface.

Medial malleolus The projection at the distal end; commonly called "ankle bone."

Articular surfaces for the fibula and the talus (a tarsal bone) There are two articular surfaces for the fibula, one at the proximal end and one at the distal end; the articular surface for the talus is at the distal end.

Interosseous border, or crest Located on the lateral side.

Fibula

Lateral malleolus Located at the distal end. The lateral "ankle bone."

Articular surfaces One for the tibia at the proximal end, and one for the tibia and one for the talus at the distal end.

Interosseous border, or crest On the medial side.

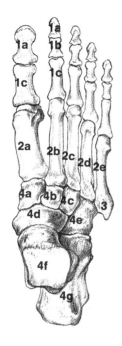

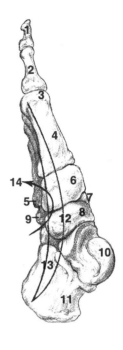

Figure 1-41
Arches of foot (right), medial aspect

1 Distal phalanx
2 Proximal phalanx
3 Head of first metatarsal
4 First metatarsal
5 Fifth metatarsal
6 Medial (first) cuneiform
7 Intermediate (second) cuneiform
8 Navicular (scaphoid)
9 Cuboid
10 Talus (astragalus)
11 Calcaneus
12 Medial longitudinal arch
13 Lateral longitudinal arch
14 Transverse arch

Figure 1-40
Bones of foot (right), upper surface

1 Phalanges
  a Distal phalanx
  b Middle phalanx
  c Proximal phalanx
2 Metatarsals
  a First metatarsal
  b Second metatarsal
  c Third metatarsal
  d Fourth metatarsal
  e Fifth metatarsal
3 Tuberosity of fifth metatarsal
4 Tarsals
  a Medial (first) cuneiform
  b Intermediate (second) cuneiform
  c Lateral (third) cuneiform
  d Navicular (scaphoid)
  e Cuboid
  f Talus (astragalus)
  g Calcaneus

Tarsals

  Calcaneus The heel bone.
  Talus, or **astragalus**
  Navicular On the medial side.
  Cuboid On the lateral side.
  Cuneiforms
    Medial, or **I**
    Intermediate, or **II**
    Lateral, or **III**

Metatarsals These are comparable to the metacarpals of the hand. Like the metacarpals, they have no names, and are numbered as follows, beginning on the medial (or large toe) side: 1st, 2nd, 3rd, 4th, 5th. (The first one in the cat is very rudimentary.)

Phalanges These are the bones of the digits, comparable to the phalanges of the fingers. Two in each large toe: proximal, distal. Three in each smaller toe: proximal, middle, distal. (The cat does not have a genuine large toe, just a rudimentary one.)

Arches of the foot

Transverse

Longitudinal

The two-way arch construction makes a highly stable base.

Articulations of the pelvic appendage

Sacroiliac joint That between the sacrum and the ilium. In the very young person this is a combination cartilaginous and synovial joint, with some characteristics of a gliding joint. It becomes wholly cartilaginous (a synchondrosis) in the adult.

Interpubic joint That between the two pubic bones; this is known as the **pubic symphysis.**

Hip joint The joint between the acetabulum of the innominate bone and the head of the femur; a ball-and-socket joint.

Knee joint That between the femur and the tibia, and between the femur and the patella. Functionally this is a hinge joint; anatomically it is made up of three joints: two condyloid joints between the condyles of the femur and the tibia, and a gliding joint between the femur and the patella.

Tibiofibular joints

  Proximal A gliding joint.
  Middle A fibrous interosseous membrane connects the shafts of the tibia and fibula; this may be classified as a syndesmosis.
  Distal A syndesmosis.

Ankle joint (talocrural) That between the talus and tibia and fibula; it is a hinge joint.

Intertarsal joints Those between the tarsal bones; all are gliding joints. (Eversion and inversion movements of the foot are produced in the intertarsal joints.)

Tarsometatarsal joints Those between the tarsal and metatarsal bones; all are gliding joints.

Intermetatarsal joints Those between the metatarsals; all are gliding joints.

Metatarsophalangeal joints Those between the metatarsals and the proximal phalanges. Generally classified as condyloid joints, but sometimes as ball-and-socket.

Interphalangeal joints Those between the phalanges; all are hinge joints.

2 DISSECTION OF THE CAT

MATERIALS

Large plastic bag (30-gallon size for a large specimen)
1½ yards unbleached muslin for each cat
1 dissecting kit for each student
Newspapers

REMOVING THE SKIN

The embalmed cat usually comes in a plastic bag, since it is obtained from a biological supply house. If the bag has not been opened, open it and drain the excess fluid from it. Place a few layers of newspaper on the table and place the bag over these. Remove the cat from the bag and lay it dorsal side up on the bag, which will serve as a working surface.

Grasp the skin at the nape of the neck and, using the scalpel or scissors, make an incision across the nape of the neck just barely through the skin,* then make a midline incision through the skin that extends forward over the skull, down the back, and about two inches onto the tail. Sever the tail at the point where the incision ends, using bone shears or saw, and discard it.

Carefully remove the skin, working from the center of the back toward the sides. Pull the skin away from the body, using the probe and/or your fingers as much as possible. Where it is necessary to use the scalpel, keep the sharp edge directed toward the skin and away from underlying muscles. As you peel the skin off, note the nerves and blood vessels coursing in segmental arrangement under the skin, within the superficial fascia.

In gross anatomy, layers or sheets of general fibrous connective tissue are called **fasciae** (singular: **fascia**). The superficial fascia, or subcutaneous connective tissue, varies in thickness and density in different areas of the

*The skin is also called the *integument*, from the Latin *integumentum*, a covering.

body, and there will be varying amounts of fat deposit. (Connective tissue filled with fat is called **adipose** tissue.) It is the superficial fascia that allows for movability of the skin.

It will be necessary to judiciously make other incisions as you proceed in the appendage areas. When you have removed the skin as far to the sides as possible, proceed to the ventral side. You may find it more convenient to make another incision along the midventral line and again dissect out toward the sides. When removing the skin from the head and face, leave the superficial nerves and blood vessels as intact as possible.

In the trunk area an extensive muscle will be removed with the skin. This muscle, which is closely bound to the underside of the skin, is the **cutaneous maximus,** or **c. trunci,** and it is not present in the human. Fibers of the cutaneous maximus merge with the **latissimus dorsi** and **pectoralis** muscles (Fig. 3-4) in the axillary region, and you must take care that parts of these latter muscles are not removed. In the neck region another cutaneous muscle, the **platysma,** will probably be removed with the skin.

Try to leave intact the **cephalic vein** (Fig. 3-1) on the dorsal side of the pectoral appendage and the **greater saphenous vein** (Fig. 3-15) on the medial side of the pelvic appendage. If it is possible, also leave intact the **lesser saphenous vein** (Fig. 3-1) on the dorsal side of the pelvic appendage.

The female cat will have varying amounts of mammary gland tissue on the ventral side of the thorax and abdomen. This should be removed, but do not damage the underlying muscles. If you are dissecting a male cat, be careful that reproductive structures in the **inguinal** and **scrotal regions** (Fig. 5-4) are not destroyed. Do *not* remove the scrotal covering of the testes.

It may be desirable to leave the skin on the appendages, especially the forearm and leg, until just before they are studied. This will prevent unnecessary drying of the tissues. It may also be wise to leave the skin on the head until you have learned about the superficial structures, so that you will know what to look for. If you do remove the skin from the appendages at this time, remove it only to the wrist and ankle joints; the remaining skin can be removed when the muscles of the region are dissected.

Clean off as much fat and other loose subcutaneous tissue as possible. If the specimen has been properly cleaned, the individual muscles should be visible. If, after thorough cleaning, the cat is covered with hairs, rinse it in *cold* water. Drain off excess water and wrap the cat in muslin and then place it in the plastic bag, which you have cleaned off. This package can then be placed in the new plastic bag for storage.* The inner plastic bag can continue to serve as a working surface as you proceed with the dissection.†

Wrap the skin and other discarded tissues in the newspaper and put them in a waste can. Clean the table surface you have used before leaving the laboratory.

---

*It will be necessary, from time to time, to dip the cat in a preservative solution (usually 2–3 percent phenol) to keep it moist and well preserved. When storing the cat, be sure that it is securely wrapped, first in muslin and then in plastic. Any part of the muslin left exposed will continuously leak moisture to the atmosphere.

†In my own laboratory, while we are working on the cat, we put the muslin (which will be wet with preservative) in the outer plastic bag and use only the inner bag as a working surface, with perhaps some newspaper under it. This helps to reduce vapor and odor from the preservative.

Origin
end of muscle
that moves the least

- insertion
end of muscle
that moves the most

Belly
center of muscle

3 MUSCULAR SYSTEM

As you proceed with the dissection of muscles, keep in mind that *dissect* means "to separate"—not to chop or dig!

A thin, tough sheet of fascia, which is a part of the deep fascia, covers the muscles and binds them together. Where this fascia is so thick that it obscures the muscle fibers, it will be necessary to remove it, but *do not* remove the **lumbodorsal fascia (fascia thoracolumbalis)** (Figs. 3-1, 3-2), which is a sheet-like tendon for muscle attachment.

Using a metal probe to break the connective tissue (use the scalpel only if necessary), separate the muscles at the cleavage lines, which should be visible if the cat has been properly cleaned. If these lines are not visible but fiber direction is apparent, pull the muscles apart slightly; this should cause the cleavage to appear. You should follow each muscle to origin and insertion as far as is practical. It will be necessary to transect superficial muscles in order to see deeper muscles, and instructions will be given when needed. In *transection* a cut is made transversely, usually at about the center of the belly of the muscle, so that enough fibers are left attached to both the origin and the insertion for later identification. Fascia under superficial muscles must be broken and removed in order to expose deeper muscles.

Dissect for deep muscles on one side only, and dissect the other side only superficially so that the superficial muscles are left intact.

In some courses the human muscles are studied in considerable detail. For these classes a number of human skeletal diagrams, which are to be used for drawing the human muscles in place, are included at the end of

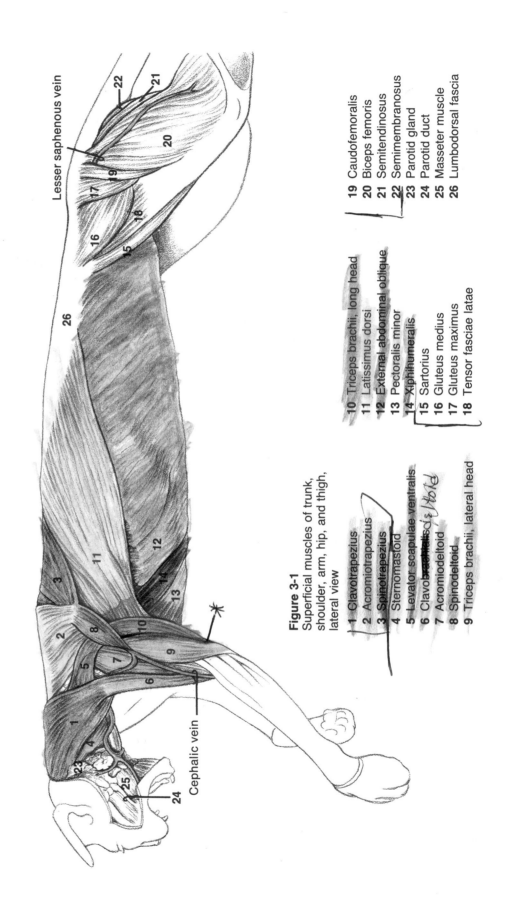

Figure 3-1
Superficial muscles of trunk, shoulder, arm, hip, and thigh, lateral view

1 Clavotrapezius
2 Acromiotrapezius
3 Spinotrapezius
4 Sternomastoid
5 Levator scapulae ventralis
6 Clavobrachialis
7 Acromiodeltoid
8 Spinodeltoid
9 Triceps brachii, lateral head
10 Triceps brachii, long head
11 Latissimus dorsi
12 External abdominal oblique
13 Pectoralis minor
14 Xiphihumeralis
15 Sartorius
16 Gluteus medius
17 Gluteus maximus
18 Tensor fasciae latae
19 Caudofemoralis
20 Biceps femoris
21 Semitendinosus
22 Semimembranosus
23 Parotid gland
24 Parotid duct
25 Masseter muscle
26 Lumbodorsal fascia

Lesser saphenous vein

Cephalic vein

this chapter. Where both right and left sides, or two or more outlines of the same side are included, use one for superficial muscles and the other(s) for deeper muscles. Indicate *origin* and *insertion* on the diagrams, and study these and *muscle fiber direction* to help you determine the kinds of movement produced by the muscles.

In using the figures provided for the study of muscles, you will note that muscles other than the ones specified for the group(s) of the particular figure are included. This is done for orientation and to show continuity. If you wish, you may indicate on the legend for the figure the other group(s) to which the muscles belong. Where a second name is given for a muscle (in parentheses), this is the name in most recent use for the cat. Nerve supply and major actions are given for most of the muscles. For more detailed information you should consult your textbook.

The easiest way to study muscles is by group, and that is the method followed here. Some muscles can be classified as belonging to more than one group. You will not observe some muscles of certain groups until you have studied other, more superficial muscles and transected them.

MUSCLE GROUPS OF THE HEAD AND NECK

Muscles of the Head and Face

Muscles of facial expression

This group is not usually studied in the cat. The muscles are essentially cutaneous muscles,* and most are removed with the skin, but you may be able to observe the fibers of a few. (Consult your textbook for descriptions of the human muscles.) These muscles are supplied by the seventh cranial nerve, the **facial** (see Fig. 8-12).

Occipitofrontalis, or **epicranius** (Fig. 3-2) Note the broad, flat tendon over the cranium between the occipital and frontal portions of the muscle. This is the **galea aponeurotica** (**aponeurosis epicranialis**). This muscle allows one to express surprise or a questioning attitude by raising the eyebrows and wrinkling the forehead.

Orbicularis oculi A sphincter muscle around the orbital opening. Contraction of this muscle allows one to squint or close the eyelids tightly.

Orbicularis oris A sphincter muscle around the oral opening (the mouth). Contraction closes the mouth.

Note the cat's well-developed muscles for moving the auricula, or pinna, of the ear.

Another superficial, cutaneous type of muscle is the **platysma**, which is a cervical or neck muscle. It is also a muscle of facial expression and is supplied by the same cranial nerve, the facial, as the other muscles of this group. It is a broad sheet arising from fascia over parts of pectoralis major and deltoideus muscles. The fibers course craniad and mediad at the side of the neck. Some fibers insert on the mandible and others terminate in

---

*You should have already observed the **cutaneous maximus** (a trunk muscle) and the **platysma** when you removed the skin. The term "cutaneous" is derived from the Latin *cutis*, meaning skin.

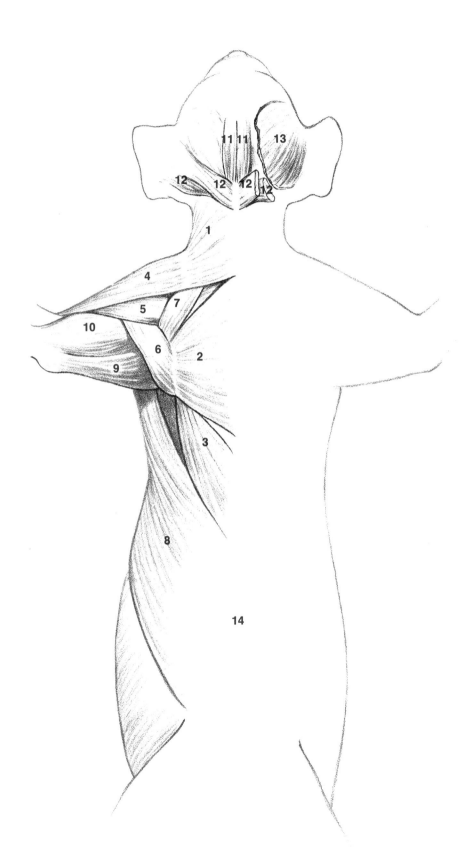

Figure 3-2
Superficial muscles of the back,
and some muscles of shoulder,
arm, and head

 1 Clavotrapezius
 2 Acromiotrapezius
 3 Spinotrapezius
 4 Clavobrachialis
 5 Acromiodeltoid
 6 Spinodeltoid
 7 Levator scapulae ventralis
 8 Latissimus dorsi
 9 Triceps brachii, long head
10 Triceps brachii, lateral head
11 Occipitofrontalis (occipital
     portion)
12 Muscles of auricula of the ear
13 Temporalis (ear muscles
     removed)
14 Lumbodorsal fascia

skin or subcutaneous tissue of the lower part of the face. In action, it draws the lower lip down and the angle of the mouth posteriorly.

Other superficial muscles are too small in the cat for satisfactory observation.

Muscles of mastication
(Figs. 3-1, 3-2, 3-4)

Note the location of the salivary glands (Figs. 3-3, 3-4). Note the location and, insofar as is practical, the attachments of the muscles described below.

Masseter A large rounded muscle in the cheek below the zygomatic arch, from which it arises. It inserts on the lateral side of the mandible. Note the duct of the parotid gland crossing the muscle (see Fig. 8-12). The masseter produces elevation of the mandible and is a powerful biting muscle. Like the other muscles of mastication, it receives a nerve supply through the mandibular branch of the fifth cranial nerve, the **trigeminal.**

Temporalis A large fan-shaped muscle at the side of the skull, that arises from the parietal bone and squamous portion of the temporal bone. In order to see this muscle, slit the galea aponeurotica and reflect it. The temporalis passes medial to the zygomatic arch and inserts on the coronoid process of the mandible. Do not follow it to insertion. The temporalis muscle, like the masseter, is a powerful elevator of the mandible and, therefore, a biting muscle.

Medial, or **internal pterygoid(eus)** This muscle cannot be observed unless a dissection is done in which the masseter muscle and a portion of the mandible are removed.* It has origin from the medial surface of the pterygoid plate of the sphenoid bone and inserts on the medial surface of the mandible, in the region of the angle. The medial pterygoid and the masseter are placed so as to suspend the mandible in a sling. The medial pterygoid is also an elevator of the mandible.

Lateral, or **external pterygoid(eus)** This muscle cannot be observed unless a deep face dissection,* or a dissection of the orbit (see Fig. 8-8), is done. It has origin from the lateral surface of the pterygoid plate and the fibers course in an essentially horizontal plane to insert on the mandible near its condyle. In the cat, which does not have a complete bony orbit, this muscle provides a floor for the orbit. The muscle acts in protrusion of the mandible, and in moving the mandible from side to side.

The following two muscles are usually placed with the hyoid muscle groups but, as they play a part in mastication and are supplied by the same cranial nerve (the trigeminal) except for the posterior belly of the digastric muscle, they are included here with the muscles of mastication.

Digastric(us) In the human, two bellies of this muscle are very evident because each ends in a tendon that is connected to the hyoid bone by a ligamentous loop. This obvious division is not present in the cat but,

*A deep face dissection is not usually attempted in undergraduate courses, and directions are not included in this manual.

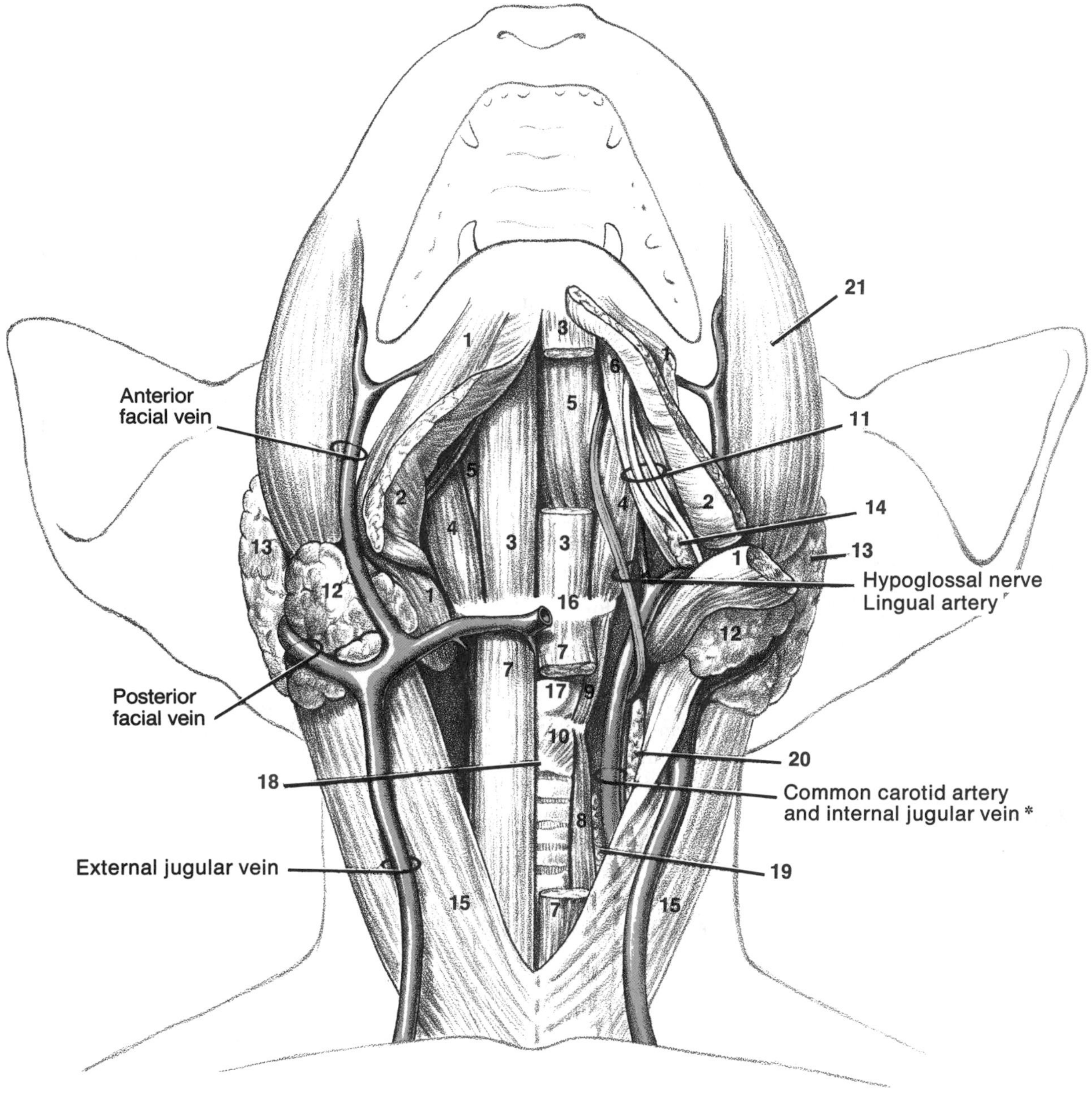

Figure 3-3
Muscles of hyoid and larynx regions, and extrinsic muscles of the tongue

1 Digastric
2 Mylohyoid (reflected)
3 Geniohyoid
4 Hyoglossus
5 Genioglossus
6 Styloglossus
7 Sternohyoid
8 Sternothyroid
9 Thyrohyoid

10 Cricothyroid
11 Ducts of submandibular and sublingual glands
12 Submandibular gland
13 Parotid gland
14 Sublingual gland
15 Sternomastoid muscle
16 Hyoid bone
17 Location of thyroid cartilage

18 Location of cricoid cartilage
19 Thyroid gland
20 Lymph node
21 Masseter muscle

* The vagus nerve, cervical sympathetic trunk and a lymphatic trunk course with these vessels.

because each end is supplied by two different cranial nerves, reference is still made to the anterior belly (supplied by the trigeminal nerve) and the posterior belly (supplied by the facial nerve). The posterior belly arises from the mastoid, or digastric groove, and the anterior belly arises from the inner border of the mandible near the symphysis menti (midline fusion area). Contraction of the muscle raises the hyoid bone (in the human) and draws it forward or backward, and it depresses the mandible in both cat and human.

Mylohyoid(eus) A thin, flat muscle with fibers that course horizontally below the mandible, deep to the digastric. Origin is from the mandible and insertion is into a median raphe and the hyoid bone. Contraction raises the hyoid bone and also the tongue.

Some Neck Muscles That Move the Head

(Figs. 3-1, 3-4, 3-6)

Sternocleidomastoid(eus) Extends from the sternum and clavicle to the mastoid process, and is supplied by the eleventh cranial nerve, the **spinal accessory,** and by ventral rami of the second and third cervical spinal nerves. Contraction of both muscles at once flexes the neck, ie, bends the head forward. Contraction of one bends the head laterad to the side of the contraction. In the cat the sternocleidomastoid consists of two separate muscles:

Sternomastoid(eus) Note the ventrolateral position of this muscle in the neck. Remove enough fascia to see the muscle clearly, but do not damage the **external jugular vein** that crosses it and passes deep between the cleidomastoid and the sternomastoid. (In the human the vein passes deep at the dorsal border of the sternocleidomastoid.)

Cleidomastoid(eus) Lateral to the sternomastoid, and deep to the ventral border of the clavotrapezius. You can observe the spinal accessory nerve piercing the cleidomastoid and coursing to the trapezius muscle.

Splenius You will not observe this muscle until you have studied and transected the trapezius. The splenius, which is deep to the trapezius, arises from the cervical ligament and inserts on the skull on a ridge between the occipital bone and the parietal bones. The human insertion is on the mastoid process and on the occipital bone just inferior to the superior nuchal line.

Contraction of both splenius muscles at the same time extends the neck, i.e., holds the head up. Singly, the splenius tilts the chin and turns the head to the side of contraction. The splenius is supplied by dorsal rami of middle and lower cervical spinal nerves.

Trapezius This muscle will be studied with another group.

Other muscles of this group will not be studied in the laboratory.

If your class is emphasizing the human muscles, draw in place, on Diagrams 1, 2, and 4 the human muscles corresponding to those you have just studied in the cat. On Diagrams 2 and 4 place the superficial muscles on one side and the deep muscles on the other.

Some Muscles of the Hyoid, Larynx, and Tongue
(Figs. 3-3, 3-4)

If the sternomastoid muscles are fused across the midline at their caudal ends, cut them apart and pull them aside in order to locate the muscles just ventral to the larynx and trachea. Muscle attachments can be determined from the names. The prefix *genio-* denotes the chin; the suffix *-glossus* denotes the tongue; *stylo-* refers to the styloid process of the temporal bone, *sterno-* to the sternum, *thyro-* and *-thyroid* to the thyroid cartilage of the larynx, *crico-* to the cricoid cartilage of the larynx, *hyo-* and *-hyoid* to the hyoid bone.

Do not disturb the bundle of blood vessels and nerves that lies laterad and dorsad of these muscles, on each side, within a thin connective tissue binding called the **carotid sheath.**

Sternohyoid(eus) A narrow, straight muscle at either side of the midline. Separate it from underlying muscles; transect and reflect on one side.

Sternothyroid(eus) Laterad and slightly dorsad of the sternohyoid. Take care not to damage the thyroid gland, which lies deep to this muscle.

Thyrohyoid(eus) Craniad of the sternothyroid and laterad of the sternohyoid.

Cricothyroid(eus) A small muscle deep to the sternohyoid.

Stylohyoid(eus) Note this tiny band crossing the outer surface of the digastric muscle to reach the hyoid bone.

Note the nerve fibers that course to the muscles. A descending branch of the twelfth cranial nerve, the **hypoglossal**, and fibers from ventral rami of upper cervical nerves (first through third in the human) supply the first three muscles listed above. The hypoglossal nerve can be observed, in company with the lingual artery, along the inferior border of the digastric muscle. The cricothyroid muscle is supplied by a branch of the tenth cranial nerve, the **vagus**, and the stylohyoid muscle by a branch of the facial nerve.

Sever the transverse vein that crosses the neck ventrally, and transect the digastric on one side, taking care not to cut underlying structures. Pass a probe under the mylohyoid to loosen it; transect and reflect to find the following muscles:

Geniohyoid(eus) Located at either side of the midline. Separate it from underlying muscles; transect and reflect on one side.

Genioglossus This lies deep to the cranial portion of the geniohyoid, and extends slightly laterad.

Hyoglossus Lies laterad of the caudal part of the geniohyoid.

Styloglossus Lies deep to the digastric, and craniad of the lingual artery and hypoglossal nerve.

The human has another slender hyoid muscle, the **omohyoid(eus)**, the superior belly of which lies at the lateral border of the sternohyoid muscle. The inferior belly, which has origin on the cranial border of the scapula, passes ventrad and slightly craniad and ends in a central tendon that is bound to the clavicle and the first rib by a process of deep cervical fascia.

The superior belly extends from this central tendon to the caudal border of the body of the hyoid bone. The attachment to the clavicle and rib gives the muscle an angular form. (The name *omohyoid* means between shoulder and hyoid.)

The geniohyoid muscle is supplied by fibers from the first cervical nerve through a branch of the hypoglossal nerve, and the omohyoid by fibers from the first three cervical nerves by way of a branch of the hypoglossal nerve. The -glossus muscles are all supplied by the hypoglossal nerve.

The muscles inserting on the hyoid bone either raise or lower it, depending on whether they are craniad or caudad of the bone. The -glossus muscles enter the tongue, where the fibers interdigitate with each other and with other fibers that are intrinsic to the tongue (ie, wholly within it). The fibers terminate in the connective tissues of the tongue. It is the interdigitation, or crisscrossing of the fibers, that makes the tongue so highly mobile.

MUSCLE GROUPS OF THE PECTORAL APPENDAGE

Muscles That Attach the Pectoral Appendage to the Vertebral Column
(Figs. 3-1, 3-2, 3-6)

Trapezius A large thin muscle, triangular in shape, which arises from the occipital bone and thoracic spines, and inserts on the pectoral girdle. This muscle and the latissimus dorsi make up the superficial layer of muscle over the back region. In the cat the trapezius has three parts:

Clavotrapezius (Cleidocervicalis) This muscle is comparable to the clavicular portion of the human muscle. It attaches to the tiny clavicle and is continuous with the clavobrachialis, which extends into the arm region.

Acromiotrapezius (Cervical trapezius) This muscle is comparable to the part of the human muscle that inserts on the acromion. Note the flat, loose tendon that connects this muscle with its counterpart on the opposite side. This connection is not as loose in the human as in the cat.

Spinotrapezius (Thoracic trapezius) This is comparable to the part of the human muscle that inserts on the scapular spine. It is superficial to the cranial border of the latissimus dorsi.

Because of its extensive origin and insertion the trapezius has various actions. Contraction of the entire muscle rotates the scapula dorsad and mediad, and also assists in abduction of the arm. Contraction of only the upper fibers pulls the scapula craniad, and are used in shrugging the shoulder. Contraction of the upper fibers on both sides at once draws the head dorsad; contraction on one side only turns the head toward the side of contraction. Contraction of the middle fibers stabilizes an upright posture, while contraction of the lower fibers draws the scapula downward. The nerve supply to the trapezius is by the spinal accessory nerve and branches from ventral rami of upper cervical nerves (the third and fourth in the human, but the first four in the cat).

Latissimus dorsi A flat, broad muscle, triangular in shape, with extensive origin in the back from spines of the last six thoracic vertebrae and all of the lumbar vertebrae, and from the sacrum and the iliac crest by aponeurosis (lumbodorsal fascia). Insertion is at the distal end of the intertubercular groove of the humerus.

L. dorsi acts as an extensor, adductor, and medial rotator of the arm. Its nerve supply is from ventral rami of lower cervical nerves by way of the brachial plexus (see p. 181).

Separate the trapezius and latissimus dorsi muscles from each other and from underlying structures by inserting the metal probe under them or by using your fingers. Transect and reflect the separate parts of the trapezius; also transect and reflect the latissimus dorsi.

Rhomboideus A rhomboid-shaped muscle situated deep to the trapezius, passing from the spines of the upper thoracic vertebrae to the vertebral border of the scapula. If two parts of this muscle can be distinguished, the cranial part is larger and could be regarded as rhomboideus minor, while the small caudal part inserting at the inferior angle of the scapula might be regarded as rhomboideus major. The cranial-caudal position is the same as in the human, but r. major is larger than r. minor in the human.

The rhomboid muscle serves to draw the scapula mediad, toward the vertebral column, and can also draw it slightly craniad. The most caudal fibers (r. major) serve to depress the lateral angle of the scapula and so assist in adduction of the arm. The nerve supply is by branches from ventral rami of cervical nerves (fifth cervical nerve only in the human).

The following three muscles all act to draw the scapula craniad, and all are supplied through ventral rami of the third and fourth cervical nerves. The human has only one of these muscles, the **levator scapulae.**

Occipitoscapularis (Rhomboideus capitis) A narrow band closely related to the rhomboideus. It arises from the occipital bone and inserts on the scapula just cranial to the insertion of the rhomboideus. It is superficial to the splenius, which can be observed at the side of the neck after superficial muscles have been transected.

Levator scapulae ventralis (Omotransversarius) Lies deep to the ventral portion of the clavotrapezius. It joins the craniolateral edge of the acromiotrapezius and inserts on the acromial part of the scapular spine. It arises from the transverse process of the atlas and the occipital bone.

To see the next muscle, pull the superior, or cranial border of the scapula from the body wall and rotate it forward; the cranial border of the muscle will appear between the inner surface of the scapula and the body wall.

Levator scapulae This is not a separate muscle, but a cranial continuation of the serratus anterior, with origin from transverse processes of the caudal five cervical vertebrae, and insertion on the vertebral border of the scapula, ventral to insertion of the rhomboideus. In the human the insertion of the levator scapulae is comparable to that of the cat, but the origin (transverse processes of the first four cervical vertebrae) is roughly comparable to the origin of the cat's levator scapulae ventralis.

Draw the human muscles in place on Diagram 2, putting the superficial muscles on one side and the deeper muscles on the other.

Muscles That Attach the Pectoral Appendage to the Ventral and Lateral Body Wall (To Sternum and Ribs)
(Figs. 3-4, 3-5, 3-7)

Pectoralis group (pectorales) The human has two muscles in this group, the **pectoralis major** and **pectoralis minor,** but the cat has four muscles, which can be placed in two subgroups as follows:

Pectoralis superficialis   The more superficial muscle is the **pectoantebrachialis** (pectoralis descendens), a narrow ribbon extending from the upper sternum to the forearm fascia. The larger muscle is the **pectoralis major** (p. transversus), which lies immediately deep to the pectoantebrachialis and can be seen below the caudal border of the latter. The fibers of both muscles run in an essentially transverse direction from sternum to humerus. In the human the pectoralis major is a larger muscle, with a more extensive origin on the clavicle, the sternum, and the costal cartilages of the sternal ribs, but the insertion is less extensive than in the cat, being confined to a small area on the humerus just distal to the greater tuberosity.

Pectoralis profundus   The larger muscle of this subgroup, **pectoralis minor** (p. ascendens), arises from the sternum caudal to the origin of pectoralis major. The fibers course diagonally craniad and laterad and pass deep to pectoralis major to insert on both the scapula and humerus. In the human the pectoralis minor is small; it arises from the ventral surfaces of the third, fourth, and fifth ribs and inserts only on the coracoid process of the scapula. In the cat the most caudal muscle, **xiphihumeralis,** arises from the xiphoid region of the sternum and inserts, along with a part of pectoralis minor and the latissimus dorsi, on the humerus.

In the human another muscle could be considered a part of this group. This is the very small, cylindrical **subclavius,** from the first rib at its junction with the costal cartilage to the inferior surface of the clavicle near its lateral end.

The pectoralis major (superficial group of the cat) flexes, adducts, and medially rotates the arm. The pectoralis minor (profundus group of the cat) draws the scapula ventrad and downward, and assists in adduction of the arm. The subclavius of the human draws the scapula ventrad and caudad. This group of muscles is supplied by nerves from the brachial plexus.

Separate and transect each of the pectoralis muscles, being extremely careful not to destroy the brachial nerve plexus (Figs. 8-15 and 8-16) and blood vessels that lie immediately deep to these muscles in the axilla. (The **axilla** is the space bounded ventrally by the pectorales, dorsally by the latissimus dorsi, teres major, and subscapularis, medially by the serratus anterior, and laterally by the arm muscles. "Armpit" is the name commonly applied to the axilla.)

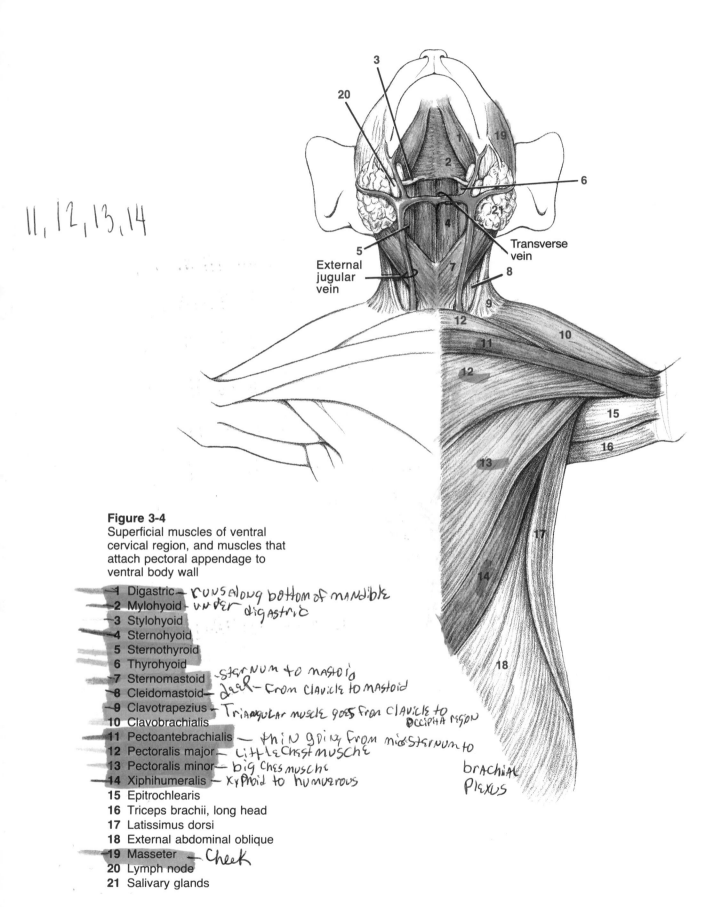

11, 12, 13, 14

Transverse
vein

External
jugular
vein

Figure 3-4
Superficial muscles of ventral
cervical region, and muscles that
attach pectoral appendage to
ventral body wall

1 Digastric — runs along bottom of mandible
2 Mylohyoid — under digastric
3 Stylohyoid
4 Sternohyoid
5 Sternothyroid
6 Thyrohyoid
7 Sternomastoid — sternum to mastoid
8 Cleidomastoid — deep - from clavicle to mastoid
9 Clavotrapezius — Triangular muscle goes from clavicle to occipita region
10 Clavobrachialis
11 Pectoantebrachialis — thin going from mid sternum to
12 Pectoralis major — little chest muscle
13 Pectoralis minor — big chest muscle
14 Xiphihumeralis — xyphoid to humerous
15 Epitrochlearis
16 Triceps brachii, long head
17 Latissimus dorsi
18 External abdominal oblique
19 Masseter — cheek
20 Lymph node
21 Salivary glands

brachial
plexus

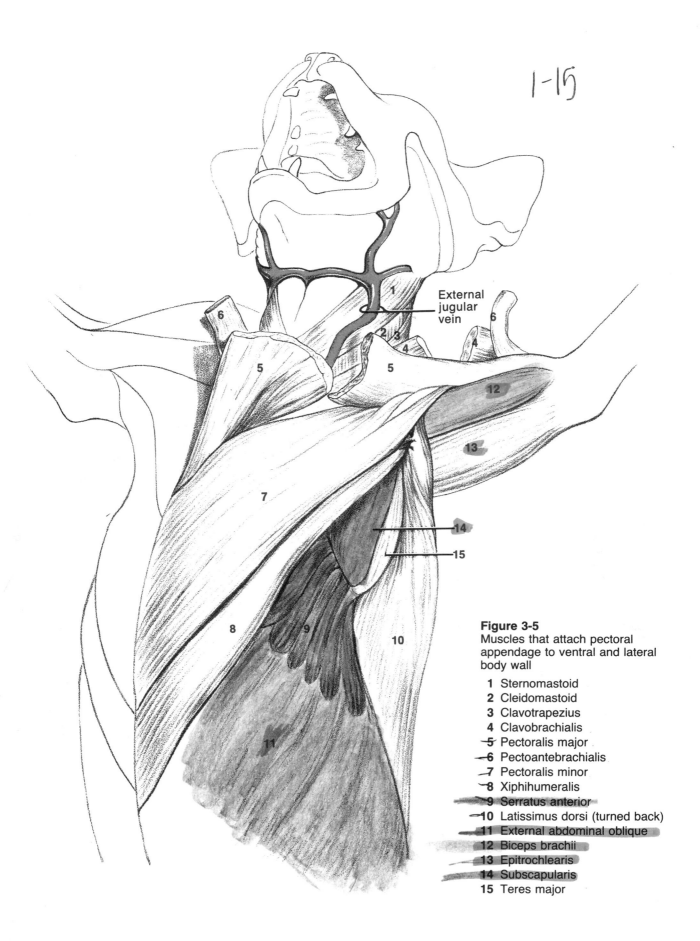

1-15

External
jugular
vein

Figure 3-5
Muscles that attach pectoral
appendage to ventral and lateral
body wall

   1 Sternomastoid
   2 Cleidomastoid
   3 Clavotrapezius
   4 Clavobrachialis
   5 Pectoralis major
   6 Pectoantebrachialis
   7 Pectoralis minor
   8 Xiphihumeralis
   9 Serratus anterior
  10 Latissimus dorsi (turned back)
  11 External abdominal oblique
  12 Biceps brachii
  13 Epitrochlearis
  14 Subscapularis
  15 Teres major

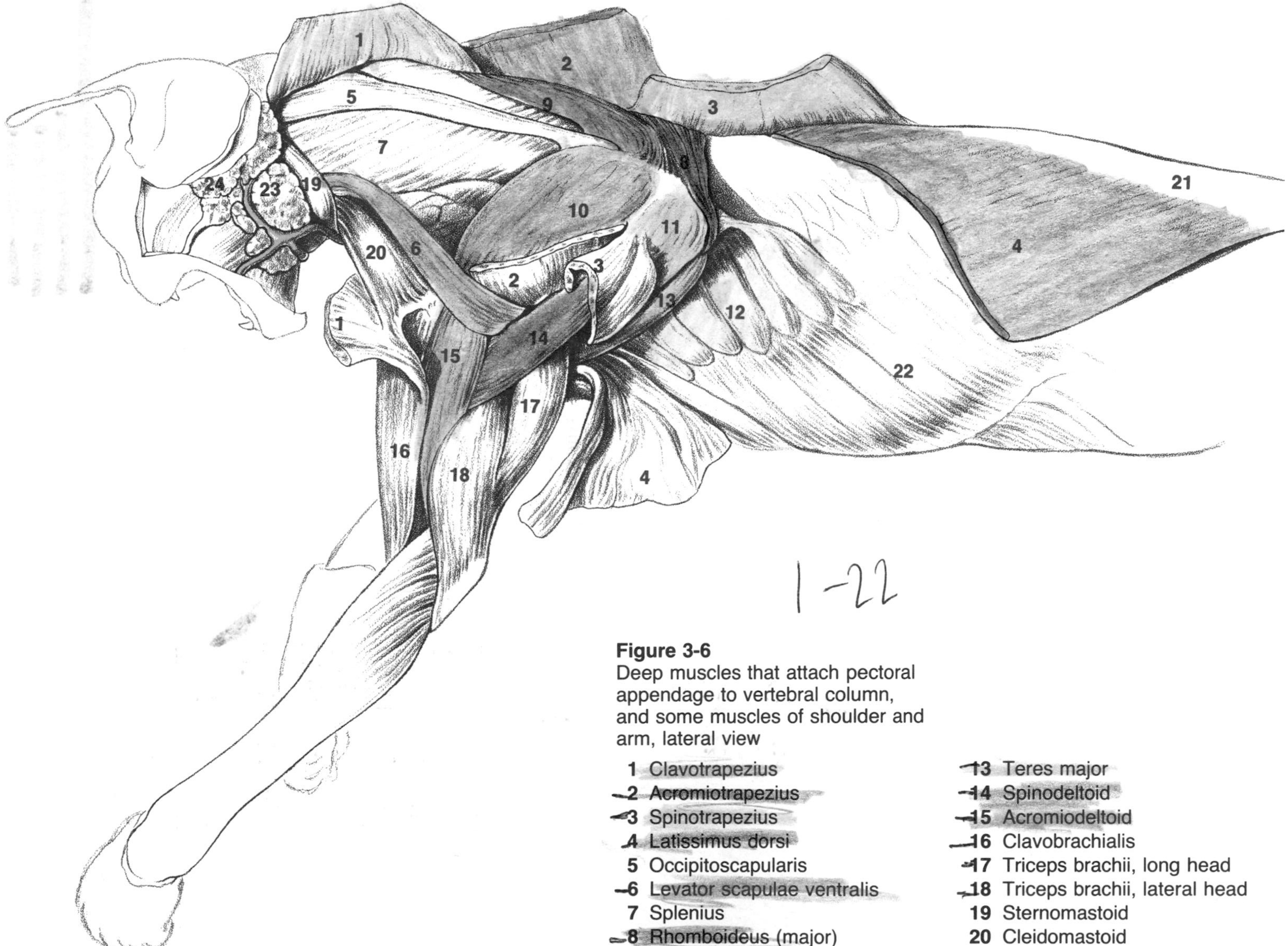

Figure 3-6
Deep muscles that attach pectoral appendage to vertebral column, and some muscles of shoulder and arm, lateral view

1 Clavotrapezius
2 Acromiotrapezius
3 Spinotrapezius
4 Latissimus dorsi
5 Occipitoscapularis
6 Levator scapulae ventralis
7 Splenius
8 Rhomboideus (major)
9 Rhomboideus (minor)
10 Supraspinatus
11 Infraspinatus
12 Serratus anterior
13 Teres major
14 Spinodeltoid
15 Acromiodeltoid
16 Clavobrachialis
17 Triceps brachii, long head
18 Triceps brachii, lateral head
19 Sternomastoid
20 Cleidomastoid
21 Lumbodorsal fascia
22 External abdominal oblique
23 Submandibular gland
24 Parotid gland

Pull the scapula away from the body wall and rotate it dorsad. Clean off enough fascia so that the next muscle can be observed.

Serratus anterior, or **ventralis,** or **magnus**   Note the serrate margin due to the origin by digitations from the upper eight or nine ribs. The digitations merge into a compact muscle that runs laterad and craniad to the vertebral border of the scapula, to insert ventral to the rhomboideus. Note the cranial continuation as the levator scapulae.

Serratus anterior draws the scapula forward and rotates it so that the glenoid fossa is moved upward, as in raising the arm. It also is important in securely anchoring the scapula to the thoracic wall. It is supplied by a nerve from the brachial plexus.

There are two posterior serratus muscles, which are not appendage muscles but are muscles of the trunk (see p. 80), and will be considered with other trunk muscles.

Draw the human muscles in place on Diagram 4, placing the superficial muscles on one side and deeper muscles on the other side.

Muscles of the Shoulder
(Figs. 3-1, 3-2, 3-6, 3-7)

Clean off as much fascia as necessary to expose muscle fibers, and then separate the muscles. All of these muscles receive a nerve supply from the brachial plexus and all act on the shoulder joint.

Deltoid(eus)   In the human this is a thick, triangular muscle with three heads of origin: clavicular, acromial, and spinous (spine of scapula). Fibers from all heads converge to insert on the deltoid tuberosity of the humerus. In the cat there are three divisions of the deltoid:

Clavobrachialis (Cleidobrachialis) This is somewhat comparable to the clavicular portion of the human muscle, and is sometimes called the *clavodeltoid* or *cleidodeltoid.* It is continuous at its proximal end (the origin) with the clavotrapezius. Some of its fibers merge with the pectoantebrachialis, which it parallels, to insert in the fascia of the forearm. Most of the insertion is with the brachialis on the ulna.

Acromiodeltoid(eus)   Comparable to the acromial portion in the human.

Spinodeltoid(eus) (Scapulodeltoid) Comparable to the spinous portion in the human. The acromiodeltoid and spinodeltoid insert together on the humerus.

As a whole, the deltoid muscle acts as an abductor of the arm but, acting independently, the clavicular portion can rotate the arm mediad and the spinous portion can rotate it laterad.

Supraspinatus Reflect the previously transected clavotrapezius and acromiotrapezius to their insertions in order to observe the supraspinatus. From its origin in the supraspinous fossa it passes ventral to the acromion, and craniad of the shoulder joint to reach the greater tuberosity of the humerus, where it inserts. The supraspinatus assists with abduction of the arm and is a weak lateral rotator.

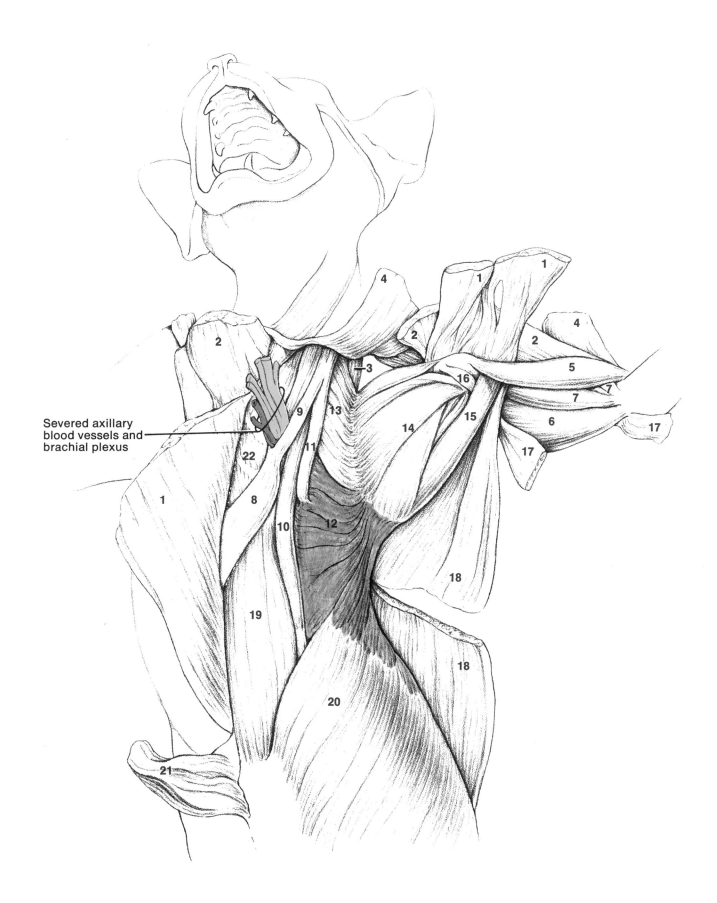

Severed axillary
blood vessels and
brachial plexus

Infraspinatus Transect the spinodeltoid and reflect. The infraspinatus can now be observed extending from its origin in the infraspinous fossa, passing deep to the spinodeltoid, to its point of insertion on the greater tuberosity of the humerus, adjacent to insertion of the supraspinatus. The infraspinatus is a lateral rotator of the arm.

Teres minor A small muscle from the axillary border of the scapula, positioned between the infraspinatus and the long head of the triceps brachii. Its insertion is just distal to that of the infraspinatus, on the greater tuberosity of the humerus. The teres minor is a lateral rotator of the arm.

Teres major From the axillary border of the scapula, caudal to the infraspinatus, this muscle passes ventral to the long head of the triceps brachii to insert with the latissimus dorsi on the humerus, along the medial margin of the intertubercular groove. The teres major is a medial rotator of the arm and also acts as an adductor and extensor.

Separate the pectorales and latissimus dorsi from the fibers of the cutaneous maximus at the axilla, and reflect to their insertions. Clean the fascia from the axilla, being very careful not to destroy the brachial nerve plexus and axillary blood vessels. The next muscle can now be observed.

Subscapularis The muscle fibers arise from the surface of the subscapular fossa and converge to a tendon that passes deep to the tendon of origin of the biceps brachii, and to the coracobrachialis, to insert on the lesser tuberosity of the humerus. (The teres major can also be observed along the axillary border of the scapula.) The subscapularis is a medial rotator of the arm and, depending on what position the arm is in, it can assist with other movements.

The tendinous insertions of the subscapularis, supraspinatus, infraspinatus, and teres minor blend with the capsular ligament of the shoulder joint to form a "rotator cuff," which reinforces the capsule and lends additional support to the shoulder joint.

Draw the human muscles in place on Diagrams 1, 2, and 4. On Diagrams 2 and 4, place the superficial and deep muscles on opposite sides to correspond to the other muscle placements you have made.

Muscles of the Arm (Upper Arm)

Before beginning the following dissection, read "Other Features of the Pectoral Appendage" (p. 60), so that you will have some notion of the landmarks and structures you should be watching for, *including the nerves that supply the muscles.*

Remove the brachial fascia from the arm muscles and separate the muscles. Take care that the **brachioradialis** (Fig. 3-8), which will be considered with muscles of the forearm, is not removed. This narrow muscle, which arises on the lateral side of the humerus and extends into the forearm, may appear to be nothing more than a strand of superficial fascia accompanied by a nerve (branch of the radial nerve).

1-22

Figure 3-7
Deep muscles that attach pectoral appendage to ventral and lateral body wall, and some muscles of shoulder and arm

1 Pectoralis minor
2 Pectoralis major
3 Splenius
4 Clavobrachialis
5 Biceps brachii
6 Triceps brachii, long head
7 Triceps brachii, medial head
8 Transversus costarum
9 Scalenus anterior
10 Scalenus medius
11 Scalenus posterior
12 Serratus anterior
13 Levator scapulae
14 Subscapularis
15 Teres major
16 Coracobrachialis
17 Epitrochlearis
18 Latissimus dorsi
19 Rectus abdominis
20 External abdominal oblique
21 Xiphihumeralis
22 Internal Intercostal

Dorsal group
(Figs. 3-6, 3-7, 3-8)

This is an extensor group. All of the muscles or their tendons cross the elbow joint and therefore produce extension of the forearm.

Triceps brachii This muscle has three heads of origin, which merge to insert by a common tendon on the olecranon process of the ulna:

Long head (middle head) Arises from the infraglenoid tubercle of the scapula. This is proportionately much larger in the cat than in the human. The long head also crosses the shoulder joint, so it acts as an extensor of the arm as well as of the forearm.

Lateral head Arises from the dorsal surface of the humerus, laterad and craniad of the radial, or spiral groove.

Medial head Arises from the dorsal surface of the humerus, medial and caudal to the radial, or spiral groove. This head has a number of divisions in the cat, which will not be studied separately.

Anconeus A very small, flat muscle arising from the dorsal surface of the lateral epicondyle of the humerus, and inserting on the olecranon process of the ulna, immediately distal to the insertion of the triceps brachii.

Ventral group
(Figs. 3-7, 3-8)

This is basically a flexor group.

Clavobrachialis (Cleidobrachialis) A division of the deltoid muscle (see p. 51).

Pectoantebrachialis (Pectoralis descendens) One of the pectorales (see p. 47).

Epitrochlearis (Tensor fasciae antebrachii) This is not present in the human. It is a flat, thin, superficial muscle on the medial side of the arm. It arises from the latissimus dorsi and is continuous distally with the antebrachial fascia (fascia of the forearm). As one of its names implies, this muscle is a tensor of the antebrachial fascia.

Remove the epitrochlearis and the pectoantebrachialis to observe the remaining muscles, which are essentially the same as those found in the arm region of the human.

Coracobrachialis In the cat this muscle is usually very small. Its origin, (coracoid process of the scapula) is the same as in the human, but its insertion on the humerus is usually more cranial in position. In the human, insertion is at about the middle of the medial surface of the humerus. The coracobrachialis is a flexor and an adductor of the arm.

Biceps brachii In the cat this muscle lies deep to the pectorales, just medial to their humeral insertions. The tendon of origin passes from a small tubercle above the glenoid fossa, through the capsule of the shoulder joint, and along the intertubercular groove. Insertion is on the bicipital tuberosity of the radius. The human biceps brachii has two heads of origin: a long head, which is comparable to the cat muscle, and a short head, which arises from the coracoid process of the scapula and joins the distal end of the long head. The insertion is the same in the human as in the cat. A fibrous band (called the **bicipital aponeurosis,** or **lacertus fibrosus**) passes obliquely mediad from the insertion to fuse with the forearm fascia. This attachment assists the muscle in its flexion action.

The biceps brachii crosses both shoulder and elbow joints, producing flexion of both the arm and forearm. Also, it is a strong supinator of the forearm due to its insertion on the radius.

Brachialis In the cat this muscle arises from the lateral side of the humerus, crosses the cubital, or antecubital fossa (ventral depression at the elbow joint), and inserts on the ventral surface of the ulna just distal to the coronoid process. In the cat the insertion of the pectoralis major separates the biceps brachii and brachialis, the brachialis being on the lateral side. In the human the brachialis arises from the distal half of the ventral surface of the humerus, so that medially it lies deep to the biceps brachii. The brachialis is a strong flexor of the forearm.

Draw the human arm muscles in place on Diagrams 6 and 7, using 6(a) and 7(a) for superficial muscles and 6(b) and 7(b) for deeper muscles.

Muscles of the Forearm and Hand

Remove the remaining skin from the forelimb. Note and remove the tough sheath of fascia (antebrachial fascia) that encases the forearm muscles. Observe the long tendons of insertion of most of the muscles, and note that tendons extending into the hand are bound down at the wrist by **transverse carpal ligaments** (dorsal and ventral, or volar). It will be necessary to sever these ligaments to follow the tendons.

It is not necessary to transect all of the muscles of the forearm. If the forearm is dissected properly, you can observe most of the deep muscles by moving aside other muscles.

Dorsal group: superficial muscles
(Fig. 3-8)

The dorsal muscles are basically extensors for the joints they cross. (Their other actions are mentioned where appropriate.) The superficial muscles are listed in order, from the radial to the ulnar side.

Brachioradialis In the cat this is a narrow ribbon, which may have been removed with the superficial fascia. It arises with the dorsal group on the lateral side of the humerus and is supplied by the same nerve (radial), but it passes ventrad along the radial side of the forearm. It inserts on the lateral side of the styloid process of the radius. The brachioradialis is a flexor of the elbow joint rather than an extensor, and is important in keeping the semipronated, flexed forearm in the carrying position.

Extensor carpi radialis longus Origin is from the lateral supracondylar ridge of the humerus, distal to the origin of the brachioradialis, and insertion is at the base of the second metacarpal.

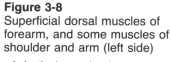

Figure 3-8
Superficial dorsal muscles of
forearm, and some muscles of
shoulder and arm (left side)

 1 Latissimus dorsi
 2 Triceps brachii, long head
 3 Triceps brachii, medial head
 4 Triceps brachii, lateral head
 5 Brachioradialis
 6 Brachialis
 7 Spinodeltoid
 8 Acromiodeltoid
 9 Pectoralis major
10 Clavobrachialis
11 Extensor carpi radialis longus
12 Extensor carpi radialis brevis
13 Extensor digitorum communis
14 Extensor digitorum lateralis
15 Extensor carpi ulnaris
16 Anconeus
17 Antebrachial fascia
18 Extensor pollicis brevis

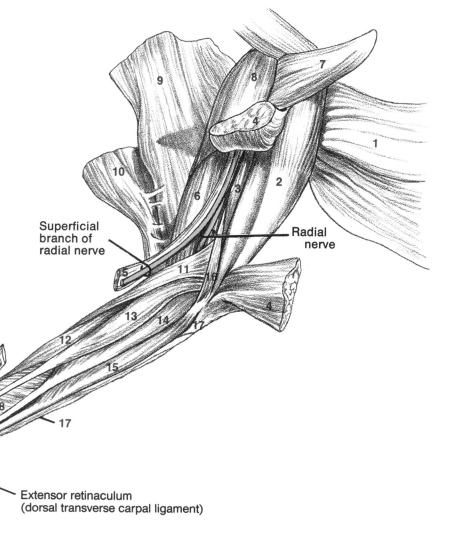

Superficial
branch of
radial nerve

Radial
nerve

Extensor retinaculum
(dorsal transverse carpal ligament)

Extensor carpi radialis brevis Origin is from the lateral epicondyle of the
humerus by a tendon common to the origins of other extensor muscles,
and insertion is at the base of the third metacarpal. The extensor carpi
radialis muscles are abductors of the hand, as well as extensors.

Extensor digitorum communis Origin is from the lateral epicondyle by
the common extensor tendon, and insertion is at the base of the middle
and distal phalanges of all digits except the thumb.

Extensor digitorum lateralis Origin is just distal to that of the extensor
digitorum communis, to which it is supplementary. The tendons pass
to the digits and join tendons of the extensor digitorum communis. The
part of this muscle that gives a tendon to the fifth digit is comparable
to the **extensor digiti quinti proprius** of the human, which is not present
in the cat as a separate muscle. The extensor digitorum lateralis is not
present in the human.

Extensor carpi ulnaris Arises from the common extensor tendon and also
from the dorsal border of the ulna. Insertion is on the medial side at
the base of the fifth metacarpal. Extensor carpi ulnaris is an adductor
of the hand, as well as an extensor.

Figure 3-9
Deep dorsal muscles of forearm
(left side)

 1 Triceps brachii, lateral head
 2 Triceps brachii, long head
 3 Brachialis
 4 Extensor carpi radialis longus
 5 Extensor digitorum communis
 6 Extensor digitorum lateralis
 8 Extensor pollicis brevis
 9 Extensor indicis proprius
10 Extensor carpi ulnaris
11 Antebrachial fascia
12 Tendons of extensores carpi
    radialis longus and brevis

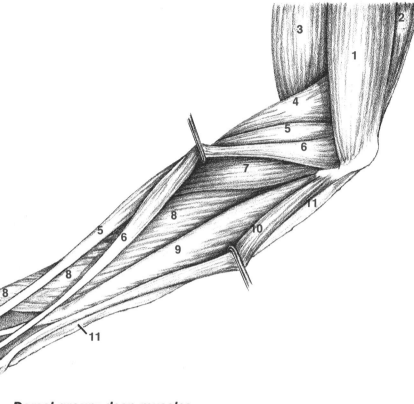

Dorsal group: deep muscles
(Fig. 3-9)

Pull aside overlying muscles to locate deep muscles.

Extensor indicis proprius Origin is from the middle of the ulna and the interosseous membrane; the tendon of insertion passes to the index finger and joins the tendon of the extensor digitorum communis. In the cat a tendon from the extensor indicis proprius goes to the thumb. This tendon corresponds to the **extensor pollicis longus** of the human, which is not present as a separate muscle in the cat. (The name indicis refers to the *index* finger; pollicis is derived from *pollex*, the Latin term for the thumb.)

Extensor pollicis brevis The origin of this muscle is more extensive in the cat than in the human, since it comes from the ulna as well as from the dorsal surface of the radius and the interosseous membrane. The muscle corresponds to both the **extensor pollicis brevis** and the **abductor pollicis longus** of the human, the latter being absent as a separate muscle in the cat. In the cat insertion is on the radial side of the base of the first metacarpal, which is the point of insertion of abductor pollicis longus in the human. In the human the extensor pollicis brevis inserts at the base of the proximal phalanx of the thumb. The extensor pollicis muscles also assist with abduction of the hand.

Supinator Origin is from the lateral epicondyle of the humerus, the radial collateral and annular ligaments, and the ulna below the radial notch. Fibers pass obliquely distad to insert on the lateral and ventral sides of the proximal third of the radius. As indicated by the name, contraction of this muscle results in a supinated forearm.

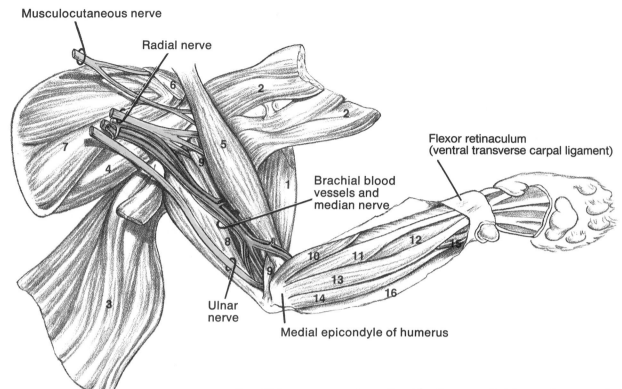

Figure 3-10
Superficial ventral muscles of forearm, and some muscles of arm (left side)

1 Pectoralis major
2 Pectoralis minor
3 Latissimus dorsi and epitrochlearis (folded)
4 Teres major
5 Biceps brachii
6 Coracobrachialis
7 Subscapularis
8 Triceps brachii, long head
9 Triceps brachii, medial head
10 Pronator teres
11 Flexor carpi radialis
12 Flexor digitorum profundus
13 Palmaris longus
14 Flexor carpi ulnaris
15 Flexor digitorum sublimis, palmaris head
16 Antebrachial fascia

It will be necessary to pull aside the superficial muscles near their origins in order to observe this muscle. (To observe the entire supinator it would be necessary to transect and reflect the overlying muscles. This is not desirable unless a good superficial dissection is done and left intact on the opposite forearm.)

Ventral group: superficial muscles
(Fig. 3-10)

The ventral muscles are basically flexors for the joints they cross—other actions are mentioned where appropriate. The muscles are listed in order, from the radial to the ulnar side, for the superficial group.

Pronator teres Origin is from the medial epicondyle of the humerus by a tendon that gives origin to other flexor muscles, and also from the medial side of the coronoid process of the ulna. It inserts on the middle third of the radius, and is a pronator of the forearm.

Flexor carpi radialis Origin is from the medial epicondyle by common flexor tendon, and insertion is at the base of the second and third metacarpals. This muscle assists the extensor carpi radialis muscles with abduction of the hand.

Palmaris longus Origin is from the medial epicondyle by common flexor tendon, and insertion is into the fascia of the palm. In the cat this muscle sends tendons to the digits, and it is larger than that in the human. (The muscle is not always present in the human. If it is present the tendon will stand out when the hand is flexed, since the tendon is not confined by the transverse carpal ligament.)

Flexor digitorum sublimis, or **superficialis** In the cat this muscle consists of two parts, each part having a muscular origin. One part arises from the tendon of the palmaris longus, and one part from the ventral surface of the distal fleshy portion of the flexor digitorum profundus. The tendons pass to all digits except the thumb, and each tendon splits to insert at each side of the middle phalanx. In the human this is a larger·muscle than in the cat and it lies deep to the palmaris longus and the flexors carpi. It has an extensive origin in the human: from the medial epicondyle by the common flexor tendon, from the medial side of the coronoid process of the ulna, and from the ventral surface of the radius distal to the insertion of the supinator.

Flexor carpi ulnaris What appears to be two muscles is actually two heads of origin: a humeral one from the medial epicondyle by the common flexor tendon, and an ulnar one from the upper two-thirds of the medial border of the ulna. Insertion is on the pisiform and hamate carpal bones, and on the base of the fifth metacarpal. This muscle assists the extensor carpi ulnaris with adduction of the hand.

Ventral group: deep muscles
(Fig. 3-11)

Transect and reflect the palmaris longus; pull aside other superficial muscles.

Flexor digitorum profundus Origin is from the upper three-fourths of the ventral and medial surfaces of the ulna and from the interosseous membrane. In the cat this muscle has five heads of origin, which will not be studied separately. The muscle sends tendons, which are bound together as they cross the wrist (under the flexor retinaculum, which should be cut), to the second, third, fourth, and fifth digits. Each tendon passes through the split tendon of the flexor digitorum sublimis and inserts at the base of the ventral surface of the distal phalanx. In the cat another tendon, which passes to the thumb, is given off, and this corresponds to the **flexor pollicis longus** of the human, which is not present as a separate muscle in the cat.

Pronator quadratus A flat, quadrilateral muscle extending across the ventral surface of approximately the distal half of the radius and ulna. Origin is on the ulna, and insertion on the radius. It is a pronator of the forearm.

In order to observe this muscle, which lies deep to the flexor digitorum profundus, you should move aside the tendon bundle of the latter, proximal to the wrist. (If it is desirable to view the entire muscle, it will be necessary to transect the flexor profundus tendon bundle proximal to the wrist and reflect the flexor profundus.)

A number of intrinsic hand muscles are present, but these will not be studied in the laboratory. Consult your textbook for descriptions of these muscles in the human.

Draw the human muscles of the forearm in place on Diagrams 6 and 7, using 6(a) and 7(a) for superficial muscles and 6(b) and 7(b) for deeper muscles.

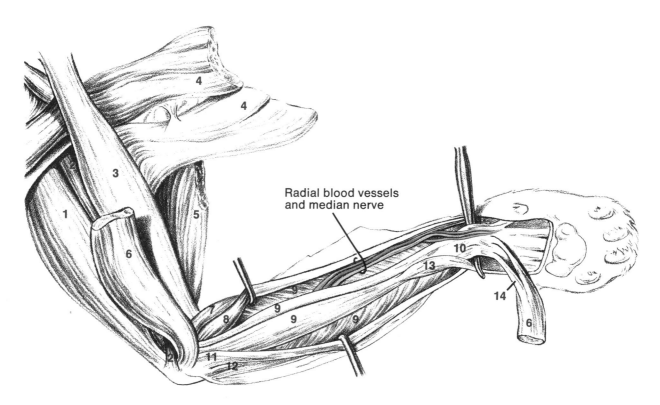

Radial blood vessels
and median nerve

Figure 3-11
Deep ventral muscles of forearm
(left side)

1 Triceps brachii, long head
2 Triceps brachii, medial head
3 Biceps brachii
4 Pectoralis minor
5 Pectoralis major
6 Palmaris longus
7 Pronator teres
8 Flexor carpi radialis
9 Flexor digitorum profundus
10 Tendon bundle of flexor
digitorum profundus
11 Flexor carpi ulnaris, humeral or
long head
12 Flexor carpi ulnaris, ulnar or
short head
13 Flexor digitorum sublimis, flexor
profundus head
14 Flexor digitorum sublimis,
palmaris head

OTHER FEATURES OF THE PECTORAL APPENDAGE
Axilla, or Axillary Fossa

This space, which contains the brachial plexus of nerves and axillary blood vessels, is described on page 47.

Cubital, or Antecubital Fossa

This fossa is the depression ventral to the bend of the elbow. It is a triangular area, whose "floor" is formed by the brachialis and supinator. The lateral border is formed by the extensor carpi radialis muscles in the cat, but by the brachioradialis in the human. The pronator teres forms the medial border. In the cat the following structures pass into or through this triangle, from lateral to medial:

Radial nerve Lies just medial to the extensor carpi radialis muscles. (In the human the radial nerve is just outside the fossa, under the brachioradialis.)

Tendon of insertion of the biceps brachii Passes between the brachialis and the supinator at their insertions.

Brachial artery (and brachial vein in the human)

Median nerve

Nerves of the Arm, Forearm, and Hand
(Figs. 3-8, 3-10, 3-11, 8-15, 8-16)

Observe the major nerves. Nerves resemble white cords, the very small ones being difficult to distinguish from strands of connective tissue. In uninjected cats nerves and arteries appear similar to one another. Identify the nerves by the muscles supplied.

Musculocutaneous Passes to the coracobrachialis, biceps brachii, and brachialis. The name of the nerve indicates that it supplies both muscle and skin.

Radial To the triceps brachii and anconeus, and all of the dorsal muscles of the forearm. You can observe this nerve where it crosses the ventral surface of the teres major and passes between parts of the medial head of the triceps brachii to the dorsal side of the arm. This nerve gives some supply to the lateral portion of the brachialis muscle in the human, and you may find this to be true in the cat. You can also locate the radial nerve superficial to the lateral surface of the brachialis by transecting the lateral head of the triceps brachii. Note the divisions into a superficial branch that passes into the superficial fascia and continues into the forearm and hand, and a deeper branch that extends to the muscles of the forearm, where it can be located passing between the supinator and the extensor carpi radialis muscles.

Ulnar This nerve courses along the medial side of the long head of the triceps brachii and passes dorsal to the medial epicondyle and into the forearm region, where it can be found deep to the flexor carpi ulnaris, which it supplies. It also supplies the ulnar head of the flexor digitorum profundus and some intrinsic hand muscles.

Median This nerve passes between the biceps brachii and the medial head of the triceps brachii on its course to the forearm. In the cat it passes through a foramen in the humerus, along with the brachial artery. (The brachial vein does not pass through this foramen.) In the forearm of the cat the median nerve is accompanied by the radial blood vessels as it courses along the ventral surface of the flexor digitorum profundus. (In the human, the radial blood vessels are more laterad and are accompanied by a branch of the radial nerve.) The median nerve supplies all ventral muscles of the forearm that are not supplied by the ulnar nerve. It also supplies some of the intrinsic hand muscles.

Blood Vessels of the Arm, Forearm, and Hand
(Figs. 3-10, 3-11, 6-5, 6-6)

Observe the major blood vessels. Arteries are white in uninjected cats, red in injected cats. Veins, which have thinner walls than arteries, usually show brown spots in uninjected cats; they are blue in injected cats. It will be practical to find only the larger vessels in uninjected cats.

Arteries

Axillary Can be located in the axilla.

Brachial A continuation of the axillary artery into the arm region. Distal to the elbow this vessel becomes known as the **radial artery.** (In the human the brachial artery divides into radial and ulnar arteries just distal to the bend of the elbow.)

Radial Passes along the ventral surface of the flexor digitorum profundus. It passes deep to the tendon of the extensor pollicis brevis and onto the dorsum of the hand. (In the human the radial blood vessels pass deep to the tendons of the abductor pollicis longus and extensores pollicis longus and brevis.)

Ulnar A small branch of the radial artery. (In the human the ulnar artery is usually larger than the radial.) The ulnar and radial arteries anastomose in the hand region to form the palmar arch (two arches in the human: superficial and deep).

Veins

Cephalic This superficial vein, which you observed when you removed the skin from the cat, courses craniad on the dorsal side of the pectoral extremity to the point at which it passes deep between the cleidomastoid and the levator scapulae ventralis, at the caudal border of the clavotrapezius. It joins a deep vein, the transverse scapular. (In the human it joins the axillary vein.)

Median cubital A branch of the cephalic vein that crosses the antecubital fossa to join the brachial vein. (In the human it joins the basilic vein, which is not present in the cat.)

Deep veins These parallel the arteries and bear the same names: brachial, radial, and ulnar.

MUSCLE GROUPS OF THE PELVIC APPENDAGE

Remove the fascia that encases the muscles of the hip and thigh so that muscle fiber direction can be determined. Take care not to remove the **iliotibial band,** or tract, which is a white, tendinous band on the lateral side of the thigh. This band is connected with the deeper fascia (fascia lata) of the thigh.

Muscles of the Hip Region

Separate the muscles of the dorsal hip region (gluteal region) at the cleavage lines.

Dorsal group: superficial muscles
Figs. 3-1, 3-12)

The dorsal hip muscles are basically extensors, abductors, and rotators. They receive a nerve supply from the lumbosacral plexus (see p. 183).

Tensor fasciae latae Located on the lateral side of the hip region, ventral in position to other muscles of this group. It arises from the outer part of the iliac crest and anterior superior iliac spine, and from a notch below the spine. The fleshy part is comparatively short, with the ventral portion lying deep to the cranial extremity of the sartorius. The fibers terminate in the iliotibial band, which inserts on the lateral condyle of the tibia. This muscle assists with abduction, medial rotation, and perhaps flexion of the thigh. Most importantly, it tightens fasciae, helping to brace the knee joint.

Caudofemoralis A small muscle arising on the caudal vertebrae and inserting by a very thin tendon on the patella. It is the most caudal of the dorsal superficial hip muscles. It is not present in the human.

Separate the caudofemoralis from the gluteus maximus and sever the tendon of insertion of the former. Reflect the muscle to its origin.

Gluteus maximus (G. superficialis) A very small muscle. It is craniad of the caudofemoralis, and its caudal border lies deep to it. Origin is from the last sacral and first caudal vertebrae and adjacent fascia; insertion is on the greater trochanter of the femur. The human muscle is much larger, with an extensive origin from the dorsal portion of the external surface of the ilium, the sacrum and coccyx, and fascia of spinal muscles; insertion is in the iliotibial band and upper dorsal surface of the femur. In the cat the caudofemoralis and gluteus maximus together are more comparable to the human muscle than the gluteus maximus alone. These two muscles act jointly in the cat as extensors and lateral rotators of the thigh in the cat.

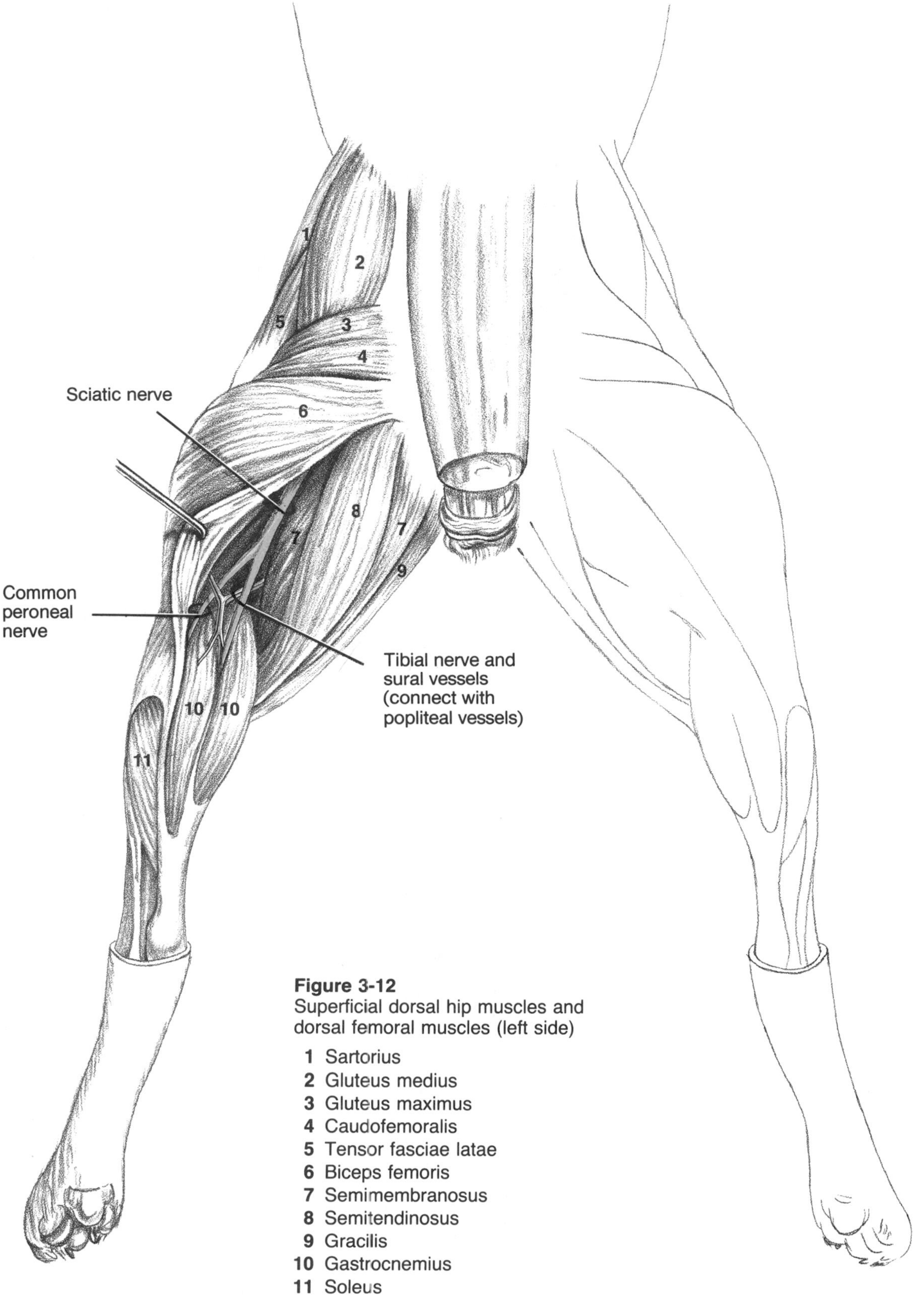

Sciatic nerve

Common
peroneal
nerve

Tibial nerve and
sural vessels
(connect with
popliteal vessels)

Figure 3-12
Superficial dorsal hip muscles and
dorsal femoral muscles (left side)

1 Sartorius
2 Gluteus medius
3 Gluteus maximus
4 Caudofemoralis
5 Tensor fasciae latae
6 Biceps femoris
7 Semimembranosus
8 Semitendinosus
9 Gracilis
10 Gastrocnemius
11 Soleus

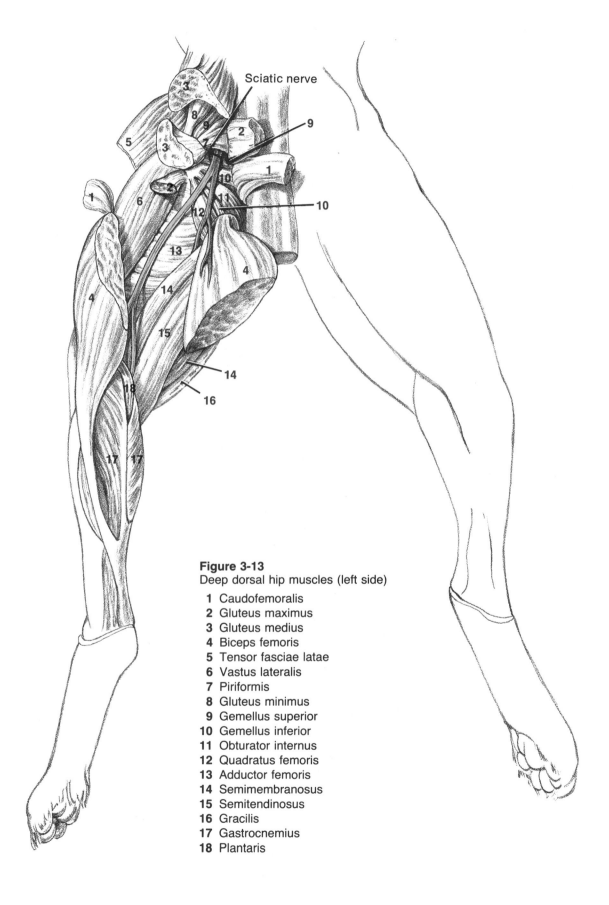

Sciatic nerve

Figure 3-13
Deep dorsal hip muscles (left side)

 1 Caudofemoralis
 2 Gluteus maximus
 3 Gluteus medius
 4 Biceps femoris
 5 Tensor fasciae latae
 6 Vastus lateralis
 7 Piriformis
 8 Gluteus minimus
 9 Gemellus superior
10 Gemellus inferior
11 Obturator internus
12 Quadratus femoris
13 Adductor femoris
14 Semimembranosus
15 Semitendinosus
16 Gracilis
17 Gastrocnemius
18 Plantaris

Sever the gluteus maximus at its insertion, and reflect it to its origin.

Gluteus medius   A thick muscle that arises from the external surface of the ilium between the anterior and inferior gluteal lies, and inserts on the greater trochanter. In the cat this muscle is larger than the gluteus maximus and will be found craniad of the latter. The gluteus medius is an abductor of the thigh. Because of its expansive origin, contraction of the anterior and posterior parts produces different results. The anterior part flexes the thigh and rotates it mediad; the posterior part extends the thigh and rotates it laterad.

Carefully separate the gluteus medius from the underlying muscles (the piriformis dorsally; the gluteus minimus ventrally) and from the tensor fasciae latae that is ventral to it. Transect and reflect.

Dorsal group: deep muscles
(Figs. 3-13, 3-14)

Piriformis   A small, flat, triangular muscle, from the ventral surface of the sacrum to the upper surface of the greater trochanter. In the human it is covered dorsally by the gluteus maximus, but in the cat by the gluteus medius. Note that the sciatic nerve passes deep to the piriformis and gluteus maximus and superficial to the gemelli muscles, the obturator internus, and the quadratus femoris. The piriformis rotates the thigh laterad, and assists with abduction and extension.

Gluteus minimus (G. profundus)   A small, somewhat pyramidal muscle lying deep to the ventral portion of the gluteus medius, with its dorsal border adjacent to the piriformis and the gemellus superior. In the human the muscle is more fan-shaped. The gluteus minimus arises from the ventral portion of the external surface of the ilium and inserts on the greater trochanter on its lateral side, slightly ventral and distal to the insertion of the gluteus medius. The actions of the gluteus minimus are similar to those of the gluteus medius.

On *one* specimen at each table, transect and reflect the piriformis in order to observe the next muscle.

Gemellus superior (G. cranialis)   A flat, triangular muscle arising, in the cat, from the dorsal border of the ischium and the ilium. Insertion is at the cranial extremity of the greater trochanter, near the insertion of the gemellus inferior and the obturator muscles. In the cat it lies deep to the piriformis, between the gluteus minimus and gemellus inferior. In the human it is caudal to the piriformis and cranial to the obturator internus, and its origin is less extensive, being only from the ischial spine.

The superior gemellus is a lateral rotator of the thigh, as is the inferior gemellus, which is described below.

Gemellus inferior (G. caudalis)   A flat, triangular muscle arising, in the cat, from the dorsal border of the ischium and the second or third caudal vertebra. It is situated just caudal to the gemellus superior. Because

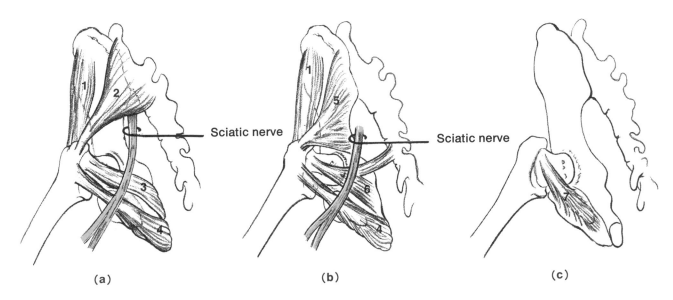

Sciatic nerve

Sciatic nerve

(a)   (b)   (c)

Figure 3-14
Cranial–caudal and dorsal–ventral alignment of deep dorsal hip muscles (left side)

(a) Shows the muscles immediately underlying the gluteus medius, gluteus maximus, caudofemoralis, and biceps femoris.

(b) The piriformis and obturator internus have been removed.

(c) The quadratus femoris has been removed to show the ventral origin of the obturator externus.

1 Gluteus minimus
2 Piriformis
3 Obturator internus
4 Quadratus femoris
5 Gemellus superior
6 Gemellus inferior
7 Obturator externus

most of it lies deep to the obturator internus, it cannot be completely observed unless the other is transected and reflected.

The gemellus inferior inserts on the tendon of the obturator internus muscle and is difficult to separate from the latter. (In the human it lies caudal to the obturator internus, arising only from the ischial tuberosity.) Maintain the obturator internus intact.

Obturator internus This muscle, lying caudal to the gemellus superior, arises from the inner surface of the ischium (from the tuberosity and the rami) (see Fig. 8-18). It passes dorsally and "curves" over the lesser sciatic notch to reach the trochanteric fossa, where it inserts by a flat tendon. The obturator internus is a lateral rotator of the thigh, and it extends and abducts the flexed thigh.

Separate the fascial connections medial to the obturator internus to expose the "curving" fibers. Lift the tendon slightly with the probe and note the attachment of fibers of the gemellus inferior on the tendon.

Quadratus femoris A short, thick muscle arising from the ischial tuberosity, caudal to the gemellus inferior. It inserts at the distal extremity of the greater trochanter, distal to insertions of the obturator and gemelli muscles. You can locate the quadratus femoris by lifting the ventral border of the proximal portion of the biceps femoris. The quadratus femoris is a lateral rotator.

Obturator externus A flat, triangular muscle, which you may not be able to observe. Near its origin, which is from the external surfaces of the inferior rami of the pubis and ischium and from the adjacent membrane over the obturator foramen, it lies deep to the adductor femoris. It passes dorsal to the hip joint to insert with the obturator internus and gemellus inferior. This muscle is deep to, and between, the obturator internus and the quadratus femoris. In some specimens it can be seen by separating the adjacent borders of the latter two muscles. The obturator externus is a lateral rotator of the thigh.

Ventral group
(Figs. 3-21, 5-1, 8-18)

The muscles of the ventral hip region (iliac region) have some origin craniad of this region, from lumbar vertebrae, and help to form the posterior, or dorsal, abdominal wall; therefore, they are also considered to be posterior abdominal muscles. Description is provided here, but actual observation will be more practical after other abdominal muscles have been studied.

Iliopsoas This is actually two muscles, the iliacus and the psoas major, which are described below. They are referred to as a single muscle because they join at their distal ends and have a common insertion on the lesser trochanter of the femur.

Iliacus The iliacus arises from the surface of the iliac fossa, and is supplied by the femoral nerve from the lumbosacral plexus. It is a flexor of the thigh.

Psoas major This muscle arises from the lumbar vertebrae. In the cat some fibers also arise from the tendons of the psoas minor (described below). Psoas major is a flexor of the thigh and of the lumbar portion of the vertebral column. Contraction on one side bends the lumbar vertebral column laterally. Both the major and minor psoas muscles are supplied by branches from ventral rami of lumbar nerves.

The distal end of the iliopsoas can be observed at this time, medial to the sartorius and the rectus femoris muscles.

Psoas minor A small muscle arising from the bodies of the last thoracic vertebrae and the first four or five lumbar vertebrae. It inserts by a thin tendon onto an eminence at the junction of the ilium and superior ramus of the pubis. In the human this muscle arises from the last thoracic vertebra and the first lumbar vertebra, but it is absent in many humans. It is larger in the cat than in the human. The psoas minor is a flexor of the pelvis and the lumbar vertebral column.

Draw the hip muscles of the human in place on Diagrams 8 and 9, placing the superficial muscles on one side and the deep muscles on the other side.

Muscles of the Thigh (Femoral Region)

Before beginning the following dissections, read "Other Features of the Pelvic Appendage" (p. 78), so you will be aware of the landmarks that you will encounter. All of the muscles in the following groups are supplied by nerves from the lumbosacral plexus (see p. 183).

Remove the remaining skin from the hindlimb, and clear away the loose fascia. Before beginning the dissection of the muscles of the thigh, observe the following superficial blood vessels:

Greater saphenous vein Courses craniad on the medial side of the leg and thigh, along with the saphenous artery and nerve, and joins the femoral vein.

Lesser saphenous vein Courses craniad on the dorsal side of the leg and thigh and passes deep between the caudofemoralis and the biceps femoris. It joins a tributary to the internal iliac vein.

Clear away as much deep fascia as necessary to expose the muscles without destroying the tendons of insertion. In clearing fat from the popliteal fossa (dorsal to the knee), take care that nerves and blood vessels within the fossa are not destroyed.

Dorsal femoral group*
(Figs. 3-12, 3-13)

These muscles, known as the "hamstring" muscles, are extensors of the hip joint and flexors of the knee joint. They all arise from the ischial tuberosity.

Biceps femoris A large muscle and the most lateral one of the group. In the cat insertion is into the dorsolateral border of the proximal third of the tibia, and into the patella. In the human insertion is onto the head of the fibula and onto the lateral side of the lateral condyle of the tibia. The biceps femoris acting alone can rotate the flexed knee laterad, although rotation of the knee is minimal. No rotation is possible with the knee joint in an extended position.

A thin ribbon of muscle, the **tenuissimus** (abductor cruris caudalis), arises near the caudofemoralis and passes along the inner surface of the biceps femoris. This muscle is not present in the human and can be disregarded.

Semimembranosus The most medial one of the group. This is a large, thick muscle which inserts on the dorsal surface of the medial condyle of the tibia in the human, but in the cat it inserts both on the femur (medial epicondyle) and on the medial surface of the tibia.

Semitendinosus The most superficial muscle of the group dorsally. It inserts into the dorsomedial border of the tibia, near its proximal end. Note the proximity of the insertions of the gracilis and sartorius muscles. Insertion is comparable in the human but extends farther ventrad, almost to the tibial tuberosity.

The above two muscles can produce a slight medial rotation of the flexed knee. They are assisted in this by the gracilis muscle of the medial femoral group, by the sartorius of the ventral femoral group, and by the popliteus muscle, which is back of the knee joint.

Ventral femoral group
(Figs. 3-15, 3-16)

Sartorius A strap-like muscle arising from the anterior superior iliac spine (origin is slightly more extensive than this in the cat). It passes diagonally across the ventral surface of the thigh, superficial to the quadriceps femoris, and inserts on the medial side of the tibia. In the human it

---

*Remember that the terms "dorsal" and "ventral" are used, in this manual, to correspond to those positions in the human.

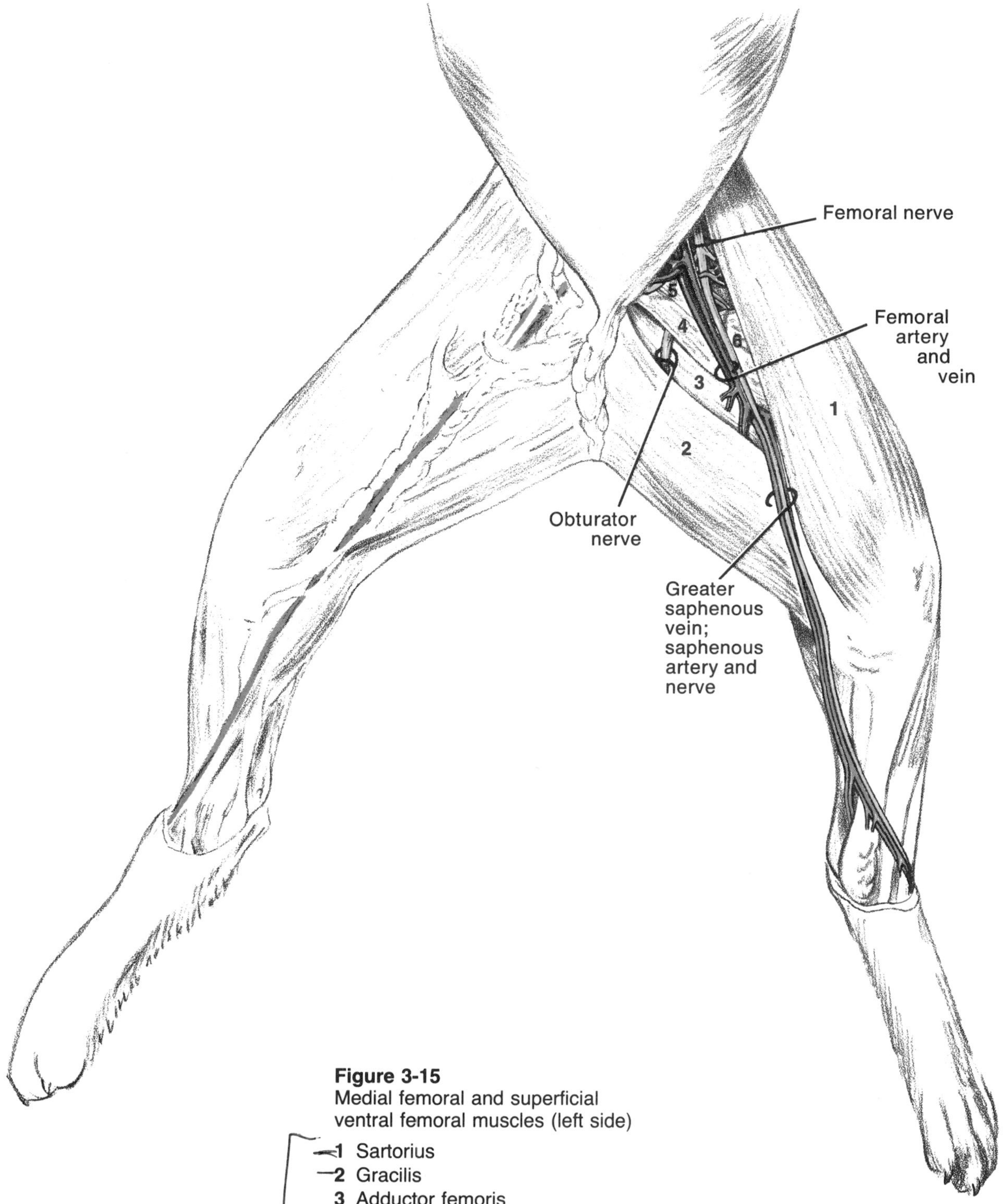

Figure 3-15
Medial femoral and superficial
ventral femoral muscles (left side)

1 Sartorius
2 Gracilis
3 Adductor femoris
4 Adductor longus
5 Pectineus
6 Vastus medialis

passes just dorsal to the medial condyle of the femur; in the cat it is much wider than in the human, and it passes medially and ventrally. Owing to this difference, it flexes the knee joint in the human, and extends it in the cat. In the cat there is also some insertion onto the patella.

The name "sartorius" is derived from the Latin term *sartor*, a tailor. The muscle is important in bringing the pelvic appendage into the tailor's sitting position (with the hip joints flexed, the knees directed laterad, and the feet crossed).

Transect and reflect the sartorius and the iliotibial band.

Quadriceps femoris This great extensor muscle of the knee joint is actually a group of muscles with a common insertion by a large tendon that extends to the patella, attaches around the patella, and passes on to the tibial tuberosity to insert. The portion of the tendon between the patella and tibia is called the **patellar ligament.** There are four divisions of this muscle:

Rectus femoris A cigar-shaped muscle that arises from the anterior inferior iliac spine and also from an area just craniad of the acetabulum. It is bordered laterally by the vastus lateralis and medially by the vastus medialis. The rectus femoris crosses the hip joint ventrally, so it is a flexor of the thigh as well as an extensor of the leg.

Vastus lateralis A large muscle arising from the lateral and dorsal surfaces of the femur (along the lateral edge of the linea aspera) and from the greater trochanter. It is ventral to the biceps femoris, and in part is also medial.

Vastus medialis A large muscle arising from the dorsal surface of the femur (along the medial edge of the linea aspera). It lies medial to the rectus femoris and lateral to the medial femoral muscles.

Vastus intermedius A flat muscle deep to the rectus femoris, and between the vastus medialis and the vastus lateralis. The origin is from the ventral surface of the femur.

Medial femoral group
(Figs. 3-15, 3-16)

This is the adductor group of the thigh. Most of these muscles can assist with flexion of the hip joint, and electromyography has shown that they can assist with medial rotation.

Gracilis A flat, wide band situated on the medial side of the thigh. Origin is from the inferior ramus of the pubis; in the cat the origin also extends to the ischium. In the cat the fibers end in a very thin, flat tendon, part of which becomes continuous with the fascia covering the distal portion of the leg. In the human insertion is near the tibial tuberosity, with the sartorius and semitendinosus.

Pectineus A small, flat muscle medial to the iliopsoas and vastus medialis, and lateral to the adductor longus. Origin is on the pubis, and insertion is on the dorsal surface of the femur just distal to the lesser trochanter (distal to the insertion of the iliopsoas).

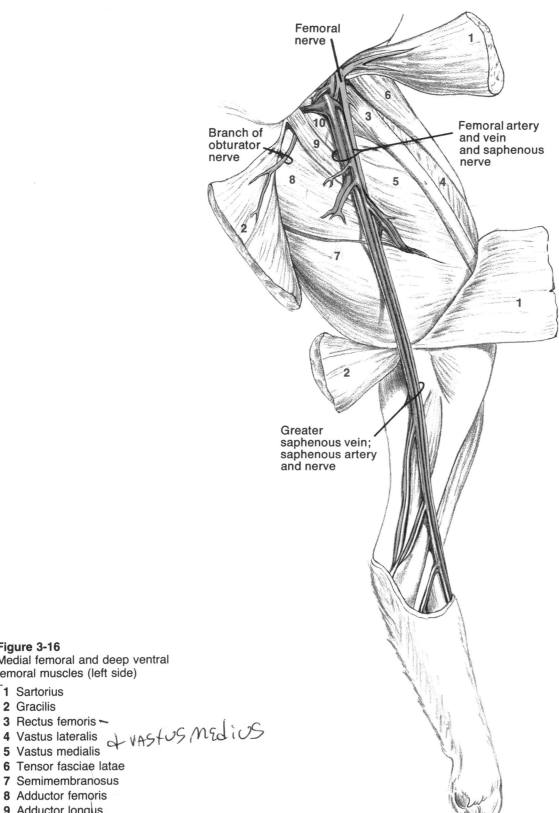

Femoral nerve

Branch of obturator nerve

Femoral artery and vein and saphenous nerve

Greater saphenous vein; saphenous artery and nerve

Figure 3-16
Medial femoral and deep ventral
femoral muscles (left side)

1 Sartorius
2 Gracilis
3 Rectus femoris
4 Vastus lateralis   & VASTUS MEDIUS
5 Vastus medialis
6 Tensor fasciae latae
7 Semimembranosus
8 Adductor femoris
9 Adductor longus
10 Pectineus

Adductor longus A thin muscle (but thicker than the pectineus) lying between the pectineus and the adductor femoris. Origin is from the cranial border of the pubis, and insertion is into the linea aspera at about the middle third. In the human the adductor longus is superficial to the **adductor brevis** and to part of the **adductor magnus,** neither of which is present as such in the cat.

Adductor femoris A large muscle lying between the adductor longus and the semimembranosus. Origin is from the inferior rami of both the pubis and the ischium; insertion is into the shaft of the femur throughout its length, along the linea aspera. This muscle corresponds to the adductores magnus and brevis of the human.

Draw the human muscles of the thigh in place on Diagrams 8, 9, and 10. On Diagrams 8 and 9 place the superficial muscles on one side and the deep muscles on the other. On Diagram 10 draw superficial muscles on 10(a), and deep muscles on 10(b). You can also place some of the hip muscles on Diagram 10, drawing the superficial muscles on 10(a), and the deep muscles on 10(b).

Muscles of the Leg and Foot

Remove the remaining skin from the foot and clean the superficial fascia from the leg and foot. Separate the muscles of the leg (crural region) by groups. It will be necessary to break some of the attachments of the femoral muscles to the fascia to leg.

Dorsal crural group: superficial muscles
(Fig. 3-17)

The muscles of this group that pass dorsal to the knee joint are flexors of that joint; those that pass dorsal to the ankle joint are extensors of that joint (ie, extension, or plantar flexion, of the foot).

Gastrocnemius Has two heads of origin: a lateral head from the lateral epicondyle of the femur, and a medial head from the medial epicondyle. The two heads unite to form a large muscle mass, which inserts by a strong tendon into the calcaneus (heel bone). The tendon is called the **calcaneal,** or **Achilles tendon.** The gastrocnemius can either flex the knee joint or extend the ankle joint; it does not perform both actions simultaneously because of the relative shortness of the muscle fibers.

Soleus Lies deep to the gastrocnemius. The cat muscle has one head of origin, the fibular head, which is deep to the lateral head of the gastrocnemius. With the gastrocnemius, the soleus inserts into the calcaneus by the calcaneal tendon. In the human there are two heads: a fibular and a tibial. The soleus is a strong extensor of the foot.

Plantaris This muscle arises from the lateral edge of the patella (and from a lateral sesamoid bone that is present in the cat). It passes between the heads of the gastrocnemius, to which it is deep, and the tendon joins the calcaneal tendon. In the human this is a small fusiform muscle of little importance, arising from the lateral epicondyle of the femur just craniad of the lateral head of the gastrocnemius.

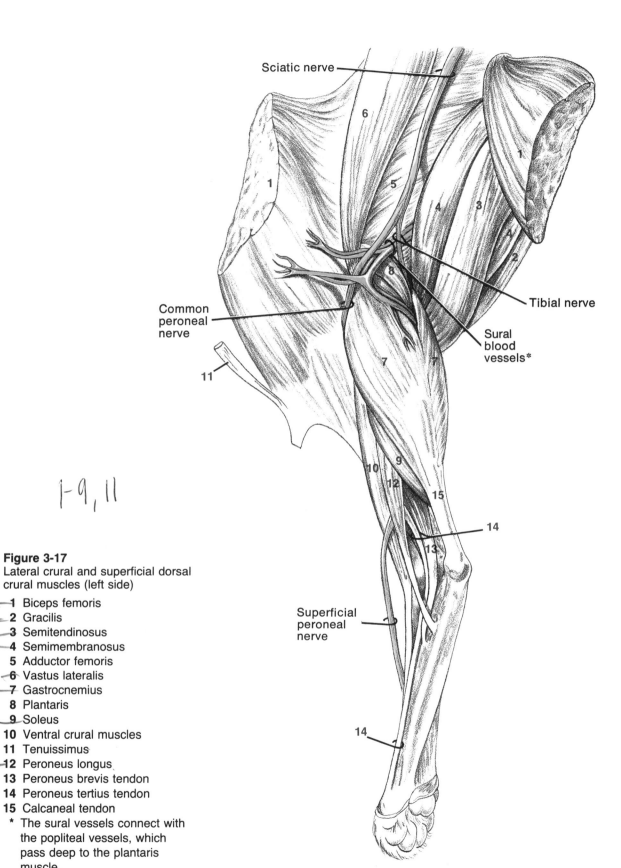

Sciatic nerve

Common
peroneal
nerve

Tibial nerve

Sural
blood
vessels*

Superficial
peroneal
nerve

1-9, 11

Figure 3-17
Lateral crural and superficial dorsal
crural muscles (left side)

  1 Biceps femoris
  2 Gracilis
  3 Semitendinosus
  4 Semimembranosus
  5 Adductor femoris
  6 Vastus lateralis
  7 Gastrocnemius
  8 Plantaris
  9 Soleus
 10 Ventral crural muscles
 11 Tenuissimus
 12 Peroneus longus
 13 Peroneus brevis tendon
 14 Peroneus tertius tendon
 15 Calcaneal tendon
  * The sural vessels connect with
    the popliteal vessels, which
    pass deep to the plantaris
    muscle.

Separate the gastrocnemius and the soleus. Sever the calcaneal tendon, taking care not to cut the tibial nerve, which is very close to the tendon. Reflect the superficial muscles toward their origins and observe the deep relationships between them; then study the next group of muscles.

Dorsal crural group: deep muscles
(Fig. 3-18)

When you separate these muscles, the best way to delimit the first three described below is to locate the tendons and follow them to the muscle mass. The tendons pass dorsal to the medial malleolus, and, listed from ventral to dorsal, they are the tibialis posterior, flexor digitorum longus, and flexor hallucis longus.

Flexor digitorum longus This muscle lies on the tibial (medial) side. Origin is from the tibia, distal to the popliteal line, and the muscle fibers end in a tendon that, in the cat, joins the tendon of the flexor hallucis longus on the plantar surface (sole) of the foot. This common tendon divides to give off tendons to all of the digits, and each tendon inserts at the base of the distal phalanx of the digit. In the human the flexor digitorum longus and flexor hallucis longus tendons are separate, with the latter going only to the big toe. The big toe is absent in the cat.

Flexor hallucis longus This muscle lies on the fibular side at its origin, which is on the distal two-thirds of the fibula and adjacent interosseous membrane. It is fairly large in the cat, and the distribution of its tendons, as noted above, is in common with the tendon distribution of the flexor digitorum longus. In the human the tendon of the flexor hallucis longus inserts at the base of the distal phalanx of the big toe (*hallex*, or *hallux*, in Latin).

Tibialis posterior (T. caudalis) A slender, fusiform muscle between the two muscles above, and deep to them. Origin is from the upper half or two-thirds of the interosseous membrane and the adjacent surfaces of the tibia and fibula. Muscle fibers converge to a tendon that passes deep to the flexor digitorum longus tendon and emerges ventral to it. Insertion is on the plantar surface of the foot.

The above three muscles are extensors of the foot. The flexor digitorum longus and flexor hallucis longus flex the digits, and the tibialis posterior inverts the foot. None of the three has origin above the knee joint, so they have no action on it.

Popliteus A triangular muscle deep to the proximal portion of the gastrocnemius. It arises just distal to the origin of the lateral head of the gastrocnemius and fans out to insert on the medial side of the dorsal surface of the tibia, proximal to the popliteal line. Contraction of the popliteus produces a very slight medial rotation of the knee. (See an anatomy textbook for a more detailed analysis of the actions of this muscle.) The popliteus forms the "floor" of the popliteal fossa (see p. 78) and is crossed dorsally by the tibial nerve and the posterior divisions of the popliteal blood vessels. The anterior tibial blood vessels pass ventral to the popliteus.

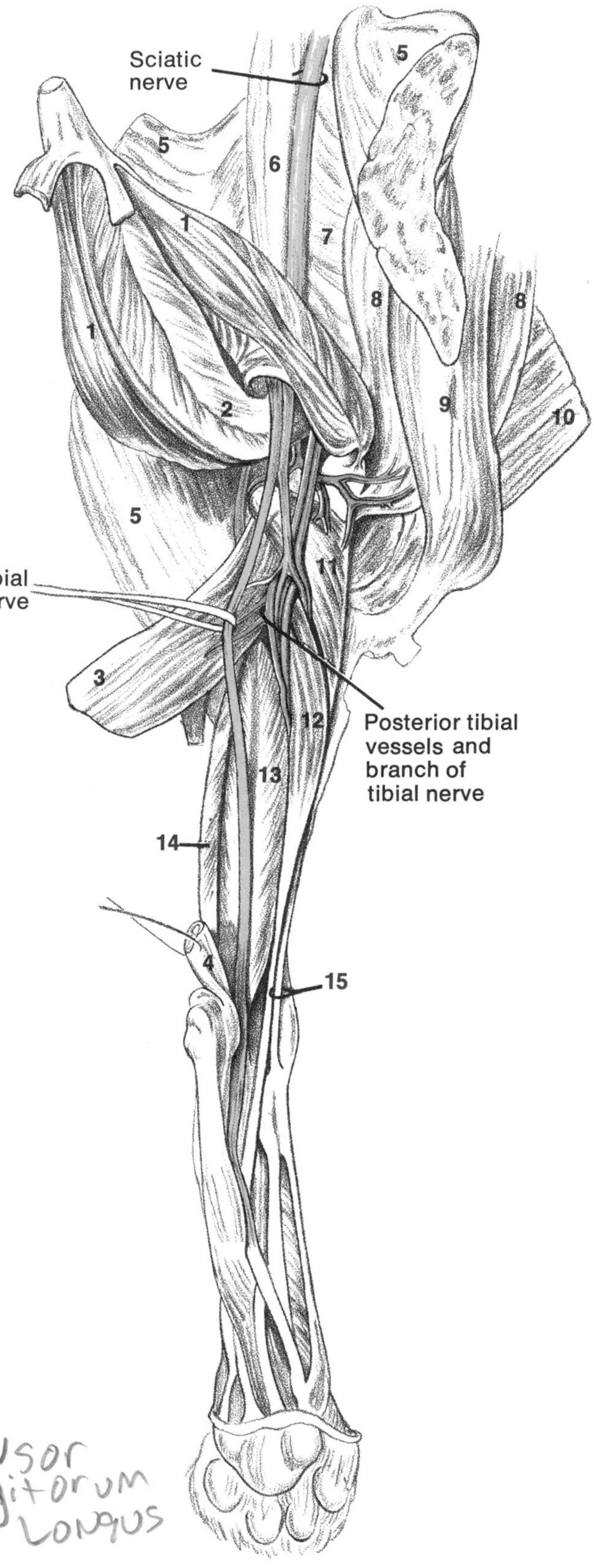

Figure 3-18
Deep dorsal crural muscles
(left side)

1 Gastrocnemius
2 Plantaris
3 Soleus
4 Calcaneal tendon
5 Biceps femoris
6 Vastus lateralis
7 Adductor femoris
8 Semimembranosus
9 Semitendinosus
10 Gracilis
11 Popliteus
12 Flexor digitorum longus
13 Flexor hallucis longus
14 Lateral crural muscles
15 Tendon of tibialis posterior

Lateral crural group
(Fig. 3-17)

Peroneus longus A slender, fusiform muscle situated at the lateral side of the leg, where it arises from the proximal half of the fibula. In the human, origin is also partly from the tibia. In the cat the tendon passes across the ventrolateral surface of the lateral malleolus and crosses the tendons of the peroneus tertius and brevis, to reach the plantar surface of the foot, where it inserts at the base of the metatarsals. In the human the tendon does not insert on all metatarsals, but only on the first, and on the first or second cuneiform.

Peroneus brevis This muscle arises from the distal half of the fibula in the cat, but from the middle third in the human. It ends in a thick tendon, which, in the cat, passes across the dorsolateral surface of the lateral malleolus in a groove, along with the tendon of the peroneus tertius. Insertion, in both cat and human, is on the lateral side of the tuberosity of the fifth metatarsal.

Peroneus tertius A slender, fusiform muscle that arises from about the middle of the lateral surface of the fibula. In the cat, its tendon crosses the dorsolateral side of the lateral malleolus in a groove, along with the tendon of the peroneus brevis, and inserts on the tendon of the extensor digitorum longus that extends to the fifth digit. In the human the peroneus tertius is a part of the ventral crural group, arising from the distal third of the ventromedial surface of the fibula. It is closely adjacent to the extensor digitorum longus and is often considered to be a part of it. The human tendon inserts at the superior surface of the base of the fifth metatarsal. (Also in the human, the peroneus tertius receives a nerve supply from the same nerve that supplies the ventral crural muscles, the deep peroneal branch of the common peroneal nerve.)

The peroneus muscles are everters and abductors of the foot, and the peroneus tertius extends and abducts the small toe. These muscles also assist with flexion and extension of the foot, but there is some variation between the cat and the human, owing to a difference in location of the tendons as they cross the ankle joint. In the human, the tendons of the peroneus longus and brevis pass behind the lateral malleolus and both muscles help to extend the foot. In the cat the tendon of the peroneus longus passes across the ventrolateral surface of the lateral malleolus, and the muscle helps to flex the foot; the tendon of the peroneus brevis passes dorsolateral and the muscle helps to extend the foot. In the human the peroneus tertius, with its tendon crossing the ankle joint ventrally, assists with flexion of the foot. In the cat the peroneus tertius tendon crosses the ankle joint in a groove on the lateral malleolus, along with the tendon of the peroneus brevis; the muscle is small and any action produced, aside from that on the small toe, is minimal.

Ventral crural group
(Fig. 3-19)

The muscles of this group are the flexors of the ankle joint (dorsiflexion of the foot). The tibialis anterior also adducts and inverts the foot, and those that send tendons to the digits extend the digits.

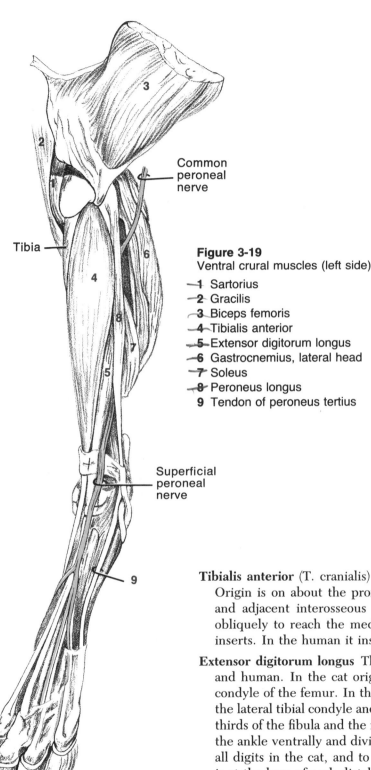

Figure 3-19
Ventral crural muscles (left side)
1 Sartorius
2 Gracilis
3 Biceps femoris
4 Tibialis anterior
5 Extensor digitorum longus
6 Gastrocnemius, lateral head
7 Soleus
8 Peroneus longus
9 Tendon of peroneus tertius

Tibialis anterior (T. cranialis) The most superficial muscle of this group. Origin is on about the proximal half of the lateral surface of the tibia and adjacent interosseous membrane. The tendon crosses the ankle obliquely to reach the medial surface of the first metatarsal, where it inserts. In the human it inserts also on the first cuneiform.

Extensor digitorum longus The origin of this muscle differs between cat and human. In the cat origin is by a flat tendon from the lateral epicondyle of the femur. In the human origin is from the lateral surface of the lateral tibial condyle and the ventromedial surface of the upper two-thirds of the fibula and the interosseous membrane. The tendon crosses the ankle ventrally and divides into four tendons that are distributed to all digits in the cat, and to all but the big toe in the human. Insertion is at the base of each distal phalanx.

In the human, as mentioned above, the peroneus tertius belongs to the ventral group. The human has another ventral crural muscle that is not present in the cat. This is the **extensor hallucis longus,** which arises on the middle half of the fibula and adjacent interosseous membrane, and passes to the base of the distal phalanx of the big toe to insert. The muscle lies between the tibilis anterior and the extensor digitorum longus.

There are a number of intrinsic muscles of the foot that will not be studied in the laboratory.

Draw the human crural muscles in place on Diagrams 10, 11, and 12. Place the lateral crural muscles on Diagram 10. The dorsal muscles should be placed on Diagram 11: the superficial muscles on 11(a) and 11(b), the deep muscles on 11(c). On Diagram 12, place the most superficial ventral muscles on 12(a) and the deeper ventral muscles on 12(b). Place as many lateral crural muscles on both 11 and 12 as you can without interfering with placement of the other muscles.

OTHER FEATURES OF THE PELVIC APPENDAGE
Femoral Triangle
(Fig. 3-15, 8-18)

Note the triangle formed by the medial border of the sartorius and the lateral border of the gracilis, with the other medial femoral muscles providing the "floor." Find, from lateral to medial, the following: femoral nerve, femoral artery, femoral vein.

Popliteal Fossa
(Figs. 3-17, 3-18)

This is the space dorsal to the knee joint at the distal end of the femur. It is bounded by the biceps femoris laterally, the semimembranosus medially, and the plantaris and the heads of the gastrocnemius distally. The "floor" is formed by the popliteus. This space contains the popliteal artery and vein, the tibial nerve, lymph nodes, connective tissue, and fat.

Nerves of the Thigh, Leg, and Foot
(Figs. 3-15, 3-16, 3-17, 3-18, 8-18)

Femoral Emerges from the psoas major and distributes to the ventral femoral muscles and to the pectineus of the medial femoral group. It gives off a cutaneous branch, the saphenous nerve, which passes distad, with the saphenous artery and greater saphenous vein, to supply the skin on the medial and ventral sides of the leg and foot.

Obturator Emerges from the obturator foramen and supplies the medial femoral muscles, except the pectineus.

Sciatic Follow this nerve from the hip, where it crosses the quadratus femoris (deep to the biceps femoris), to its division into the tibial and common peroneal nerves. Note the branches to the dorsal femoral muscles, all of which the sciatic nerve supplies. The sciatic nerve of the cat gives off a cutaneous branch, the sural nerve, which supplies the skin on the dorsal and lateral sides of the leg and foot. Sural nerves in the human branch from the tibial and common peroneal nerves.

Tibial Passes through the popliteal fossa, courses deep to the superficial dorsal crural muscles, and passes into the foot. This nerve supplies all of the dorsal crural muscles and some of the intrinsic foot muscles.

Common peroneal Divides into a deep peroneal branch, which supplies the ventral crural muscles, and a superficial peroneal branch, which supplies the lateral crural muscles. These branches continue into the foot to supply intrinsic muscles.

Blood Vessels of the Thigh, Leg, and Foot
(Figs. 3-15, 3-16, 3-17, 3-18, 8-18)

Arteries

Femoral Locate the artery in the femoral triangle between the femoral vein and nerve. Follow it to the dorsal side of the femur and to the popliteal fossa, where it is called the **popliteal artery.**

Note the saphenous artery branching from the femoral and coursing with the saphenous nerve and greater saphenous vein. (On uninjected cats it will not be practical to follow the blood vessels distal to the popliteal fossa.)

Popliteal This artery gives off branches to adjacent tissues, including a sural artery to the biceps femoris and gastrocnemius muscles, and divides into a posterior tibial and an anterior tibial artery at the distal end of the popliteal fossa. (There is more than one sural branch in the human.)

Posterior tibial Ramifies in the dorsal crural muscles in the cat; in the human it passes distad on the leg and into the foot region. It also gives off a branch in the human, the **peroneal** artery, which courses down the fibular side of the leg to ramify in the heel region.

Anterior tibial Passes ventral to the popliteus muscle, pierces the interosseous membrane between the tibia and fibula and continues distad, on the ventral side of the leg, to the foot.

All of the arteries give off various branches throughout their course and give branches that anastomose (unite) in the foot to form a plantar arch. The plantar arch sends branches to the digits.

Veins

Superficial veins The saphenous veins have been described in the section on muscles of the thigh (see p. 67).

Deep veins These parallel the arteries and bear the same names: femoral, popliteal, anterior tibial, and posterior tibial (and peroneal in the human).

There are many communicating trunks between the superficial veins, and between the superficial veins and the deep veins.

When dissection of the deep muscles of the appendages has been completed on one side of the specimen, a superficial dissection should be made on the opposite side. Separate the superficial muscles as much as possible without destroying them. Do not transect any of them.

MUSCLES OF THE TRUNK

Thoracic Region
(Figs. 3-20, 3-21)

Scaleni group Find the scaleni muscles on the side of the specimen on which the pectorales and latissimus dorsi have been transected. These small muscles, deep to the pectorales, arise from the cervical vertebrae and insert on the upper ribs. There are three of these: an anterior, a middle, and a posterior muscle. In the cat the middle muscle is larger than the others and extends farther caudad, crossing the serratus anterior near the ventral border of the latter. Because the scaleni assist in elevating the ribs and pulling them outward, they are synergists to other muscles of inspiration. They also assist in bending the trunk ventrad and laterad. The scaleni muscles receive a nerve supply through branches of ventral rami of the lower cervical nerves.

Medial to the scaleni is the cranial portion of the rectus abdominis and the small **transversus costarum** (also known as **sternalis,** or **rectus thoracis**) superficial to it. Loosen the cranial portion of the rectus abdominis caudal to the transversus costarum, and transect. Reflect to the cranial border of the external abdominal oblique in order to observe the intercostal muscles.

External intercostals (Intercostales externi) There are eleven pairs in the human; twelve pairs in the cat. Note the direction of the fibers forward and downward, from the caudal border of one rib to the cranial border of the next rib caudad. Note that the external intercostals do not reach the sternum, and that the internal intercostals can be seen in the interval. The external intercostals are elevators of the ribs and are therefore muscles of inspiration.

Internal intercostals (Intercostales interni) These are immediately deep to the external intercostals, and they equal them in number. Note that the fibers course at approximately right angles to the fiber direction of the external muscles, as they pass from the cranial border of the rib to the caudal border of the next rib craniad. There is some disagreement on what the function of these muscles is, but they are probably depressors of the ribs and therefore muscles of expiration.

Both the external and internal intercostal muscles receive a nerve supply through intercostal nerves (ventral rami of thoracic nerves).

The space between ribs, which is occupied by the intercostal muscles, is called the **intercostal space.** Insert a probe in an intercostal space between the ventral border of the external intercostal and the underlying internal intercostal, to determine the separation.

Diaphragm This muscle will not be observed until the internal systems are studied, and description and directions will be included in the chapter on respiratory and digestive systems. When the diaphragm contracts, the ribs are pulled outward, and the thorax is thus expanded; the diaphragm is therefore a muscle of inspiration (see p. 119).

Look beneath the transected latissimus dorsi and spinotrapezius muscles and the rhomboideus to observe the posterior serratus muscles.

Serratus posterior superior (S. dorsalis cranialis) A comparatively small muscle that can be found on the dorsal side of the trunk, deep to the rhomboideus and latissimus dorsi. In the cat this muscle extends from the upper eight or nine ribs to the cervical and thoracic spines, where it attaches by aponeurosis. In the human it is less extensive, arising from the last two cervical vertebrae and the first two or three thoracic vertebrae, and inserting on the upper ribs. When the muscle contracts it elevates the ribs and is therefore a synergist in inspiration.

Serratus posterior inferior (S. dorsalis caudalis) A small muscle extending from the last four or five ribs to the lumbar spines, where it attaches by aponeurosis. In the human origin is on the last two thoracic vertebrae and the first two lumbar vertebrae, and insertion is on the last four ribs. This muscle assists in elongating the thorax and is a synergist in inspiration.

Note the continuity of the two posterior serratus muscles in the cat; for this reason the two muscles are often considered to be one muscle in the cat, serratus posterior, or dorsalis, with cranial and caudal portions.

Both of the posterior serratus muscles receive a nerve supply through intercostal nerves, as do some other muscles of the thorax that will not be studied in the laboratory at this time. These are as follows:

Levatores costarum These are small muscles, thought to aid with inspiration, from the transverse processes of thoracic vertebrae and the seventh cervical vertebra to the angles of the ribs. A single muscle extends from the transverse process of its origin to the external surface of the rib caudad of that process. These muscles will not be dissected.

Transversus thoracis This muscle has origin from the inner surface of the sternum and inserts on the inner surfaces of ribs (the second or third through the seventh, eighth, or ninth). It probably assists in forced expiration, and it can be observed when the thorax is opened.

The human has another group of very small muscles, the **subcostales,** which arise from the inner surfaces of one rib, near its angle, and insert on the second or third rib below. Usually, they are well developed only in the lower thoracic region. The fiber direction is the same as that of the internal intercostal muscles, and they have similar actions.

Abdominal Region
(Fig. 3-20)

Note the midventral fascial line on the abdomen. This line, called the **linea alba,** is formed by a fusion of the aponeuroses of the ventrolateral abdominal muscles.

Rectus abdominis A straight muscle extending from the pubis to the sternum on each side of the linea alba. It extends farther craniad in the cat than in the human and does not have the tendinous bands that pass transversely, at intervals, across the muscle in the human. The origin and insertion are reversible in their actions; that is, when either end of the muscle is fixed, the opposite end can accomplish certain movements of flexion of the trunk. It also tenses the ventral abdominal wall and assists in compressing the abdominal viscera. The nerve supply is through the lower six or seven intercostal nerves.

About 90 percent of humans have a very small triangular muscle, the **pyramidalis,** at the caudal end of the abdomen and ventral to the rectus abdominis. It arises from the pubic bone and inserts in the linea alba, of which it is a tensor. It receives a nerve supply from the twelfth intercostal nerve.

External abdominal oblique (Obliquus abdominis externus) A sheet of muscle in the ventral and lateral abdominal wall. It arises from the external surfaces of the lower nine or ten ribs in the cat (the lower eight in the human), with the cranial portion interdigitating with the origin of the serratus anterior. Insertion is on the iliac crest and by an aponeurosis which fuses with that of the opposite side to help form the linea alba. Note the general forward and downward direction of the muscle fibers, and compare with the external intercostal fiber direction.

Loosen the cranial border of the external abdominal oblique with a probe, and begin transecting the muscle in a caudal direction, approximately one-half inch laterad of the rectus abdominis. Carefully separate the muscle from the underlying muscles throughout its extent, transecting as the separation proceeds, and reflect.

Internal abdominal oblique (Obliquus abdominis internus) Lies immediately beneath the external abdominal oblique. This is also a sheet of fibers that arise from the iliac crest, the inguinal ligament (between the anterior superior iliac spine and the pubis), and the lumbodorsal fascia. Note that, in general, the fibers course at approximately right angles to those of the external abdominal oblique; compare with the internal intercostal fiber direction. The muscle fibers do not extend as far mediad in the cat as they do in the human. Insertion is by aponeurosis to the midline (linea alba). In the human there is also insertion on the costal cartilages of the lower ribs.

The cranial two-thirds of the aponeurosis of the internal abdominal oblique splits at the lateral border of the rectus abdominis, creating a dorsal leaf that passes dorsal, and a ventral leaf that passes ventral, to the rectus abdominis. The dorsal and ventral leaves unite at the medial border of the rectus abdominis to help form the linea alba. The tendinous leaves of the internal abdominal oblique and the aponeuroses of the external abdominal oblique and the transversus abdominis muscles form a sheath (rectus sheath) around the rectus abdominis muscle. The caudal third of the aponeurosis of the internal abdominal oblique does not split, and it passes ventral to the rectus abdominis.

Carefully separate the internal abdominal oblique from the underlying transversus abdominis. Because the two muscle fiber sheets and their aponeuroses are quite thin and closely applied to each other, they are difficult to separate. As the separation proceeds, transect the internal abdominal oblique in such a way that some muscle fibers will be left attached to the aponeurotic tendon. Note the relationships of the tendon with the rectus abdominis.

Transversus abdominis The innermost one of the ventrolateral abdominal muscles, arising from the lower costal cartilages, the lumbodorsal fascia, the iliac crest, and the inguinal ligament. Insertion is by aponeurosis into the linea alba. Note the general transverse direction of the fibers.

The differing fiber directions of the three abdominal muscles described above allow them to furnish much greater support for abdominal viscera than would occur if all fiber direction was the same. Contraction of the muscles compresses the abdominal viscera and aids in expulsion movements (defecation, micturition, and forced expiration). They also assist with flexion and rotation of the vertebral column.

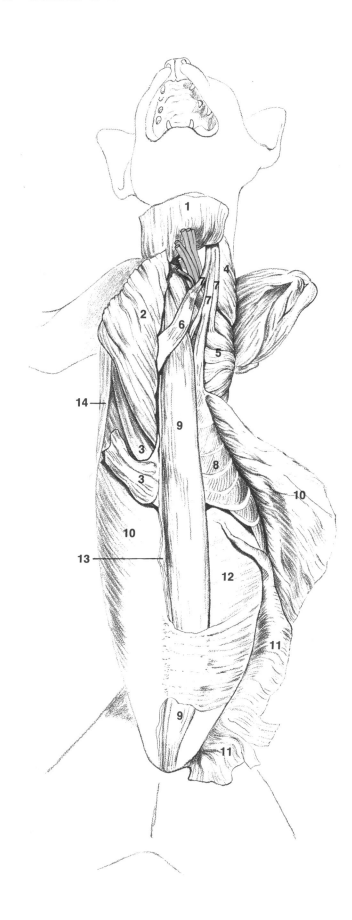

8-12

Figure 3-20
Muscles of abdomen and thorax,
ventral view

1 Pectoralis major
2 Pectoralis minor
3 Xiphihumeralis
4 Levator scapulae
5 Serratus anterior
6 Transversus costarum
7 Scaleni muscles
8 External intercostal
9 Rectus abdominis
10 External abdominal oblique
11 Internal abdominal oblique
12 Transversus abdominis
13 Cut edge of aponeuroses of
    abdominal muscles
14 Latissimus dorsi

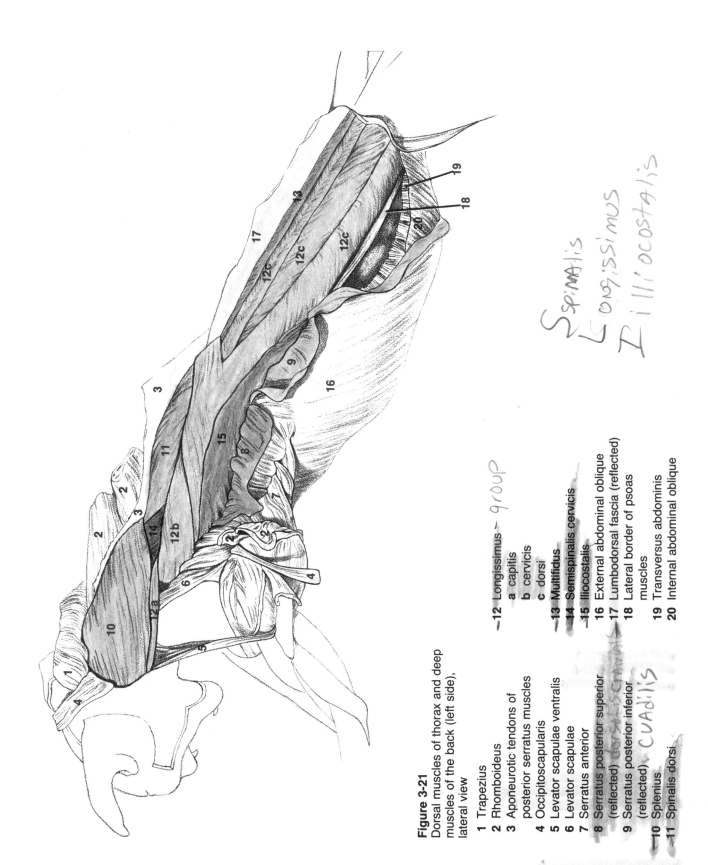

Figure 3-21
Dorsal muscles of thorax and deep muscles of the back (left side), lateral view

1 Trapezius
2 Rhomboideus
3 Aponeurotic tendons of
    posterior serratus muscles
4 Occipitoscapularis
5 Levator scapulae ventralis
6 Levator scapulae
7 Serratus anterior
8 Serratus posterior superior
    (reflected) *dorsibcranial*
9 Serratus posterior inferior
    (reflected) *cuadilis*
10 Splenius
11 Spinalis dorsi
12 Longissimus → *group*
   a capitis
   b cervicis
   c dorsi
13 Multifidus
14 Semispinalis cervicis
15 Iliocostalis
16 External abdominal oblique
17 Lumbodorsal fascia (reflected)
18 Lateral border of psoas
    muscles
19 Transversus abdominis
20 Internal abdominal oblique

Spinalis
Longissimus
Illiocostalis

Observe the aponeurosis of the transversus abdominis and the dorsal division of the aponeurosis of the internal abdominal oblique in their cranial two thirds as they pass dorsal to the rectus abdominis. Note that in approximately the caudal third, all of the aponeuroses of the ventrolateral abdominal muscles pass ventrad of the rectus abdominis, so that the caudal portion of this muscle has only the transversalis fascia (a thin layer of connective tissue) and the parietal peritoneum between it and the peritoneal cavity.

The nerve supply to the external and internal abdominal oblique and the transversus abdominis is from the lower five intercostal nerves and from branches of the ventral ramus of the first lumbar nerve (the first three lumbar nerves in the cat).

Quadratus lumborum This is a small muscle, in the cat, arising from the last two thoracic vertebrae and the last rib. It lies against the ventral surfaces of the transverse processes of the lumbar vertebrae and attaches to them; it inserts caudally on the anterior inferior iliac spine. In the human this muscle helps to form the dorsal abdominal wall and is therefore a posterior, or dorsal, abdominal muscle. Its origin in the human is from the dorsal part of the iliac crest and adjacent ligaments, and its insertion is on the transverse processes of the upper lumbar vertebrae and the inferior border of the last rib. Q. lumborum assists in flexing the lumbar portion of the vertebral column toward the side of the contracting muscle. It receives a nerve supply from ventral rami of the last thoracic and first lumbar spinal nerves. You will not observe this muscle until you study the deep back muscles.

Paying careful attention to muscle fiber direction, place the human abdominal muscles on Diagrams 4 and 5. The four sides, two in each diagram, provide ample room for drawing these muscles in place. Good judgment in arranging these to best advantage will make studying them easier.

Back Region
(Fig. 3-21)

The deep muscles of the back are epaxial muscles. The other muscles studied in the cat, with the exception of the splenius, are hypaxial muscles. Epaxial muscles are extensors of the vertebral column and are important postural muscles. They all receive a nerve supply from dorsal rami of the spinal nerves.

Reflect the latissimus dorsi and the lumbodorsal fascia to the vertebral spines, and locate the caudal portion of the first muscle described below.

Extensor dorsi communis A large muscle mass, on each side of the vertebral column, extending from the sacrum and ilium to the skull. This muscle mass is comparable to the **sacrospinalis,** or **erector spinae,** of the human. The caudal portion of the muscle arises, by a strong aponeurosis, from the iliac crest and median sacral crest, and from the lumbar vertebrae and last two thoracic vertebrae. As it extends craniad, it gives off fibers that insert on the ribs and the more cranial transverse and spinous processes. At each segmental level it receives new fibers of origin before relinquishing fibers of insertion. The intersegmental arrangement of this muscle mass provides for considerable flexibility in movement of the vertebral column.

The extensor dorsi communis is divided longitudinally into three columns. From lateral to medial these are: the **iliocostalis**, the **longissimus**, and the **spinalis.** The longissimus extends to the skull and different names are used according to the location; these are the **longissimus dorsi, cervicis, and capitis.** The spinalis is usually called **spinalis dorsi.** Some fibers of the spinalis dorsi pass cranially to join a neck muscle in the cat that is called **biventer cervicis** (the spinalis capitis of the human). In the human all three columns of the sacrospinalis have three parts, which are, from lateral to medial and caudal to cranial, as follows: iliocostalis lumborum, dorsi, and cervicis; longissimus dorsi, cervicis, and capitis; spinalis dorsi, cervicis, and capitis.

The muscle columns in the thoracic and cervical regions will not be studied closely, because it is difficult to dissect them.

Multifidus This muscle extends the length of the vertebral column, adjacent to the spinous processes. In the cat it can be observed in the lumbar region, medial to the longissimus dorsi: in the human it is covered by the sacrospinalis in this region. In the cat its cranial portions are known as the **semispinalis,** which is a separate muscle in the human.

Semispinalis This muscle lies deep to the splenius. It has two parts in the cat, the semispinalis cervicis and capitis. The medial portions of the two parts form the biventer cervicis and the lateral portions form the complexus (the semispinalis capitis of the human). In the human the semispinalis has three parts: dorsi, cervicis, and capitis.

There are other small muscles in the back that assist in moving the vertebrae, but it is beyond the scope of this manual to include these.

Sever the connection of the ventrolateral abdominal muscles with the lumbodorsal fascia. Clear away the connective tissue, being *very careful* not to disturb the contents of the abdominal cavity or to destroy the parietal peritoneum bounding the peritoneal cavity. Separate and identify the lateral borders of the posterior abdominal muscles, which lie ventral to the caudal part of the extensor dorsi communis. These are, from ventral to dorsal: psoas minor, psoas major, quadratus lumborum. The iliacus will be found dorsal to the caudal portion of the psoas major, which it joins to form the iliopsoas. These muscles can be more closely observed when the contents of the abdominal cavity are studied.

Draw the human deep back muscles in place on Diagram 3.

HUMAN SKELETAL DIAGRAMS FOR MUSCLE PLACEMENT

The following pages of diagrams are included for students who wish to emphasize origin, insertion, and action of human muscles.

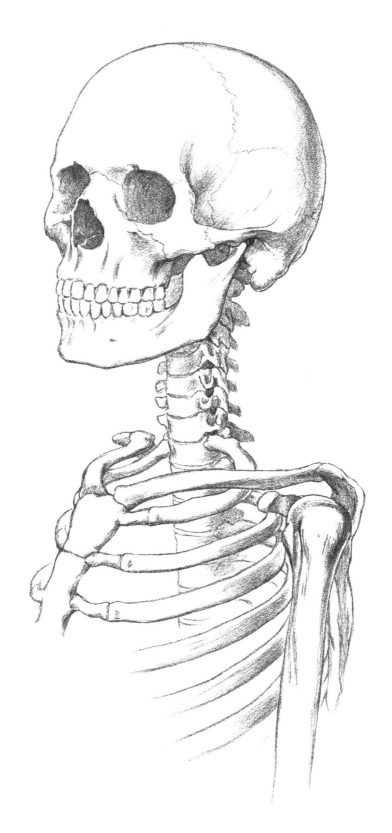

Diagram 1
Head, neck, and cranial portion of
chest region, ventrolateral view

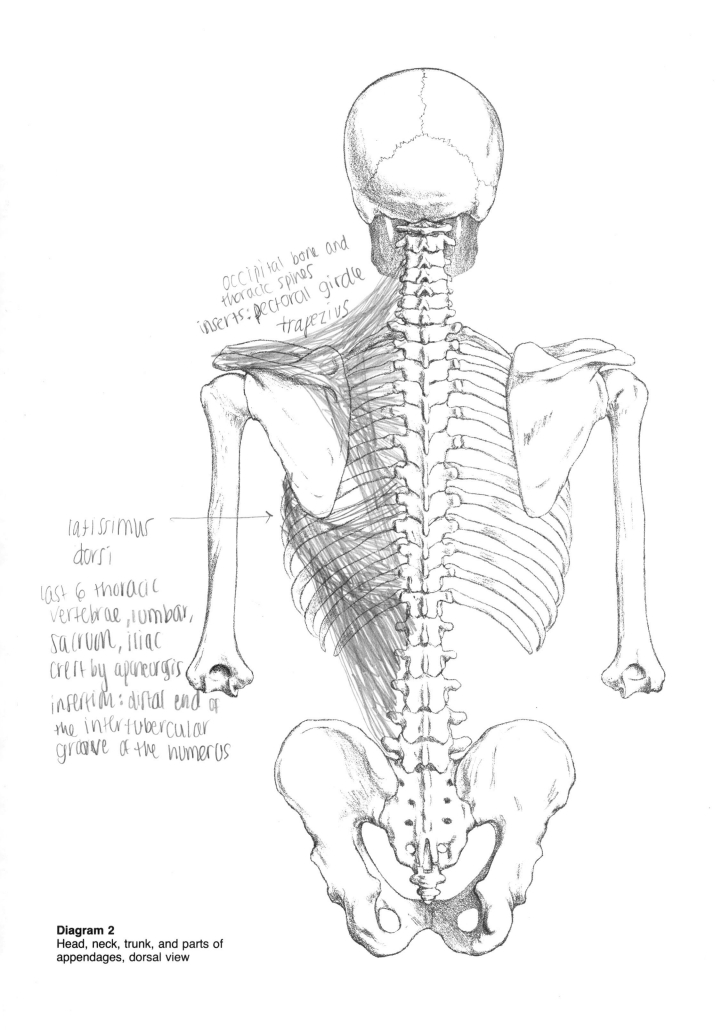

occipital bone and
thoracic spines
inserts: pectoral girdle
trapezius

latissimus
dorsi

last 6 thoracic
vertebrae, lumbar,
sacrum, iliac
crest by aponeurosis
insertion: distal end of
the intertubercular
groove of the humerus

Diagram 2
Head, neck, trunk, and parts of
appendages, dorsal view

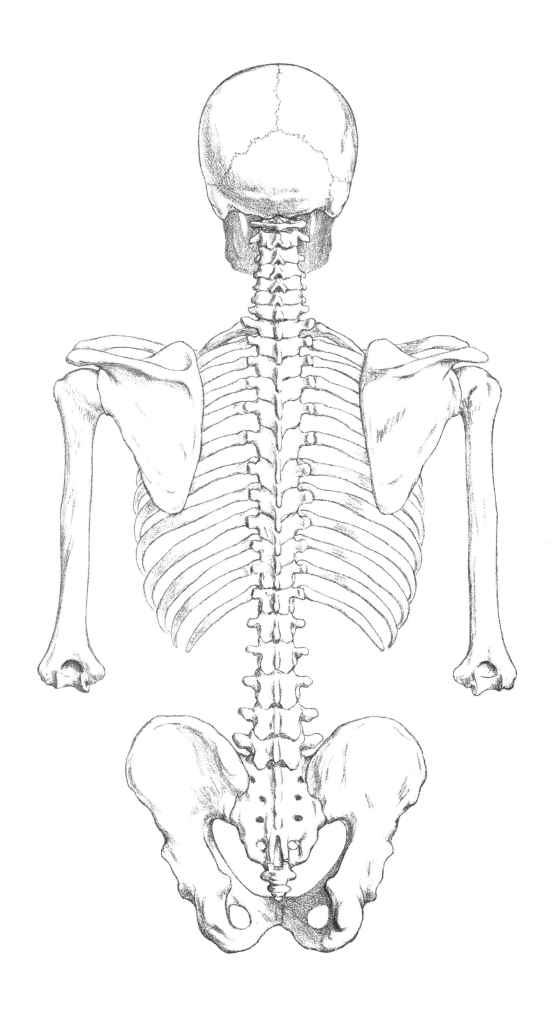

Diagram 3
Same as 2

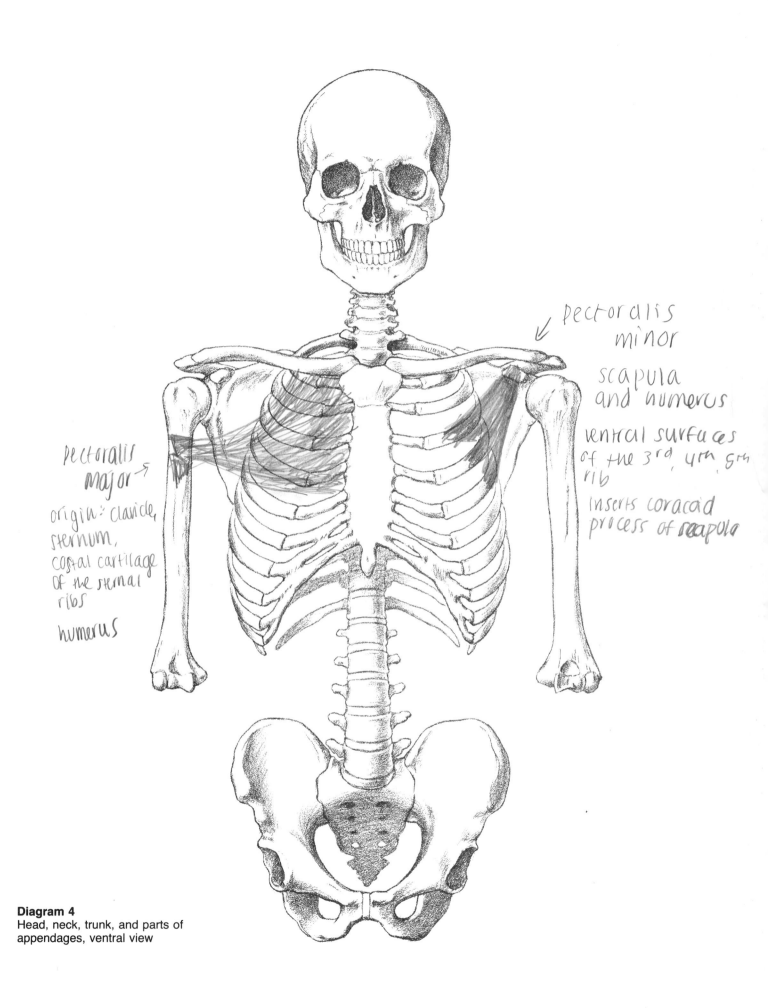

pectoralis
minor

scapula
and humerus

ventral surfaces
of the 3rd, 4th, 5th
rib

inserts coracoid
process of scapula

pectoralis
major

origin: clavicle,
sternum,
costal cartilage
of the sternal
ribs

humerus

Diagram 4
Head, neck, trunk, and parts of
appendages, ventral view

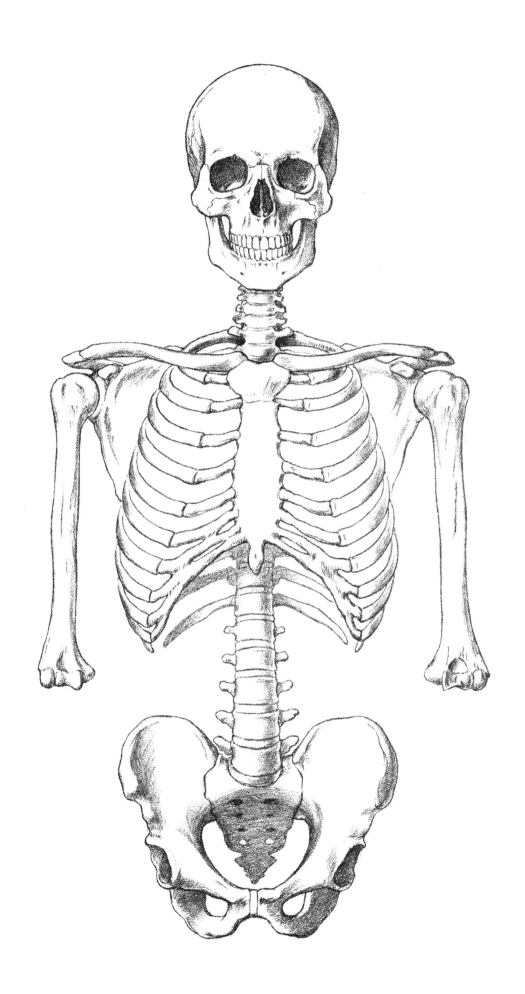

Diagram 5
Same as 4

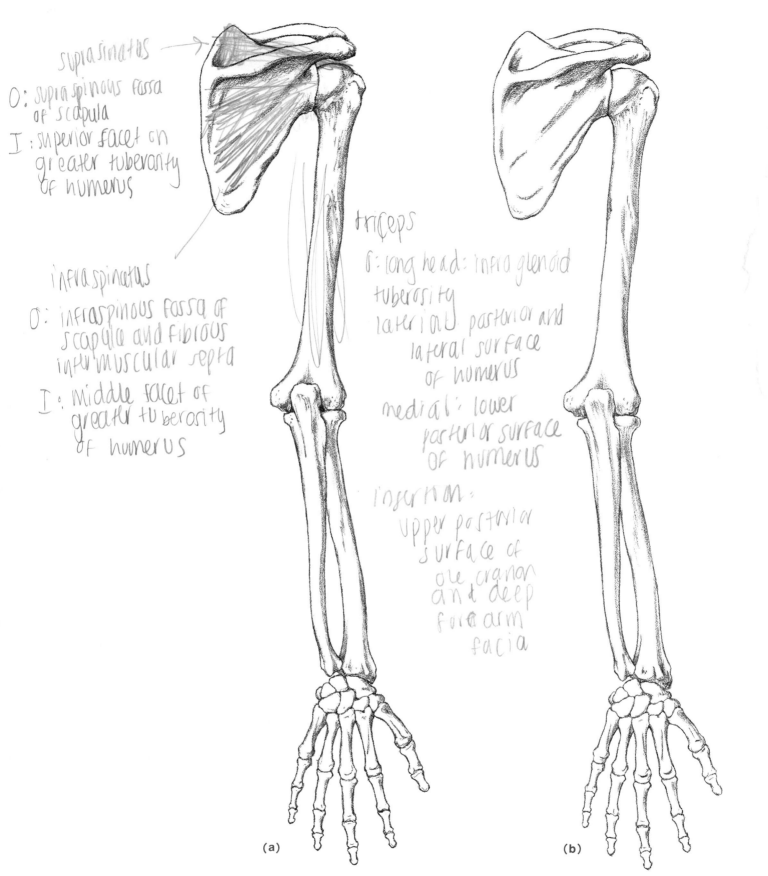

supraspinatus

O: supraspinous fossa
of scapula

I: superior facet on
greater tuberosity
of humerus

infraspinatus

O: infraspinous fossa of
scapula and fibrous
intermuscular septa

I: middle facet of
greater tuberosity
of humerus

triceps

O: long head: infraglenoid
tuberosity
lateral: posterior and
lateral surface
of humerus
medial: lower
posterior surface
of humerus

insertion:
upper posterior
surface of
olecranon
and deep
forearm
facia

(a)

(b)

Diagram 6
Shoulder, arm, forearm, and hand,
dorsal view

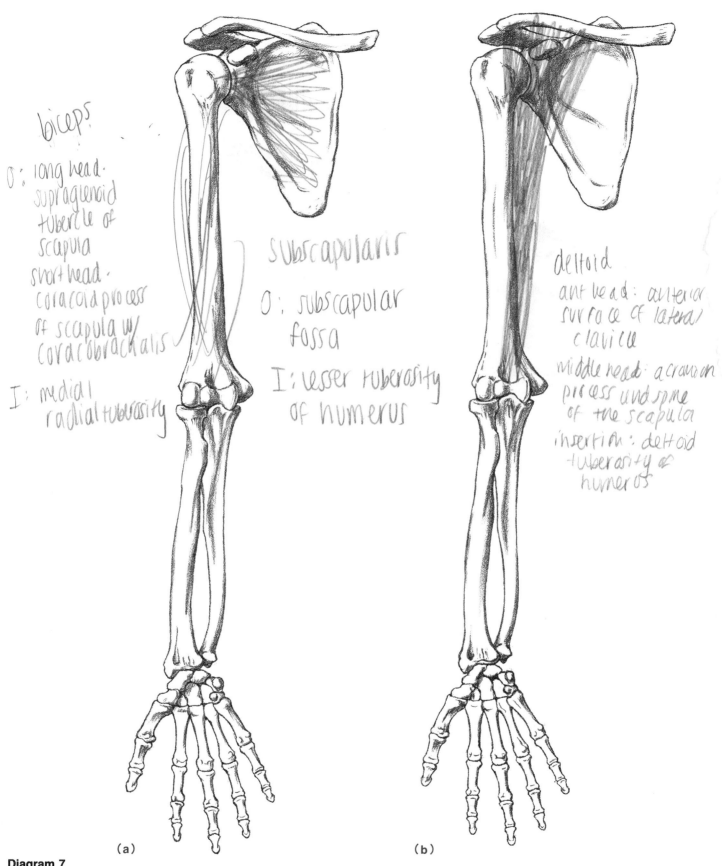

biceps

O: long head:
supraglenoid
tubercle of
scapula
short head:
coracoid process
of scapula w/
coracobrachialis

I: medial
radial tuberosity

subscapularis

O: subscapular
fossa

I: lesser tuberosity
of humerus

deltoid
ant head: anterior
surface of lateral
clavicle

middle head: acromion
process and spine
of the scapula

insertion: deltoid
tuberosity of
humerus

(a)   (b)

Diagram 7
Shoulder, arm, forearm, and hand,
ventral view

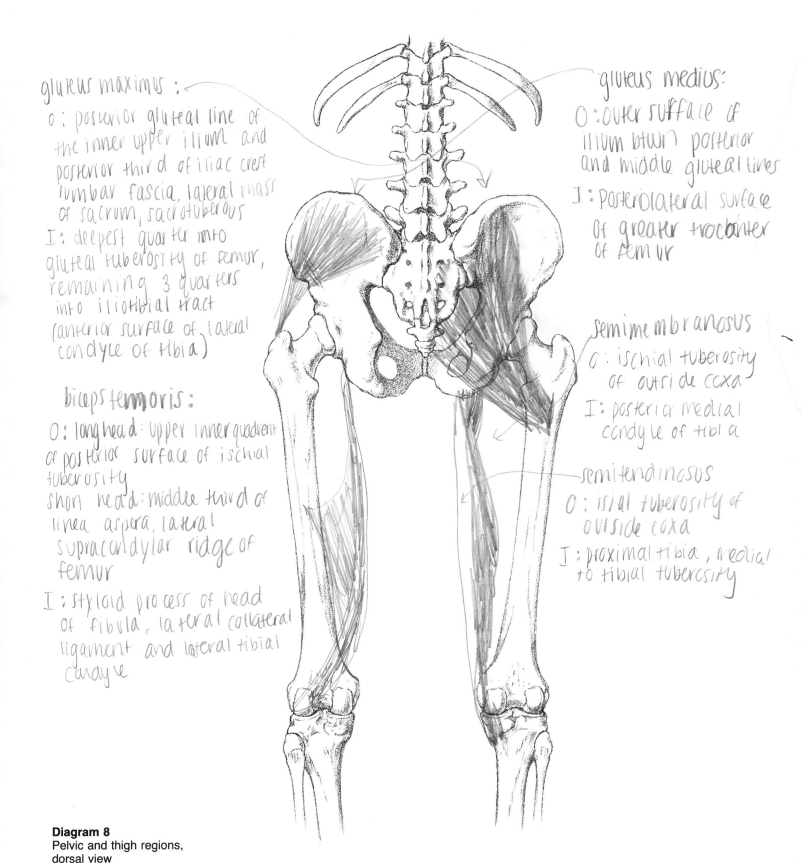

gluteus maximus :

O: posterior gluteal line of
the inner upper ilium and
posterior third of iliac crest
lumbar fascia, lateral mass
of sacrum, sacrotuberous
I: deepest quarter into
gluteal tuberosity of femur,
remaining 3 quarters
into iliotibial tract
(anterior surface of lateral
condyle of tibia)

biceps femoris :

O: long head: upper inner quadrant
of posterior surface of ischial
tuberosity
short head: middle third of
linea aspera, lateral
supracondylar ridge of
femur
I: styloid process of head
of fibula, lateral collateral
ligament and lateral tibial
condyle

gluteus medius:

O: outer surface of
ilium btwn posterior
and middle gluteal lines
I: posterolateral surface
of greater trochanter
of femur

semimembranosus

O: ischial tuberosity
of outside coxa
I: posterior medial
condyle of tibia

semitendinosus

O: isial tuberosity of
outside coxa
I: proximal tibia, medial
to tibial tuberosity

Diagram 8
Pelvic and thigh regions,
dorsal view

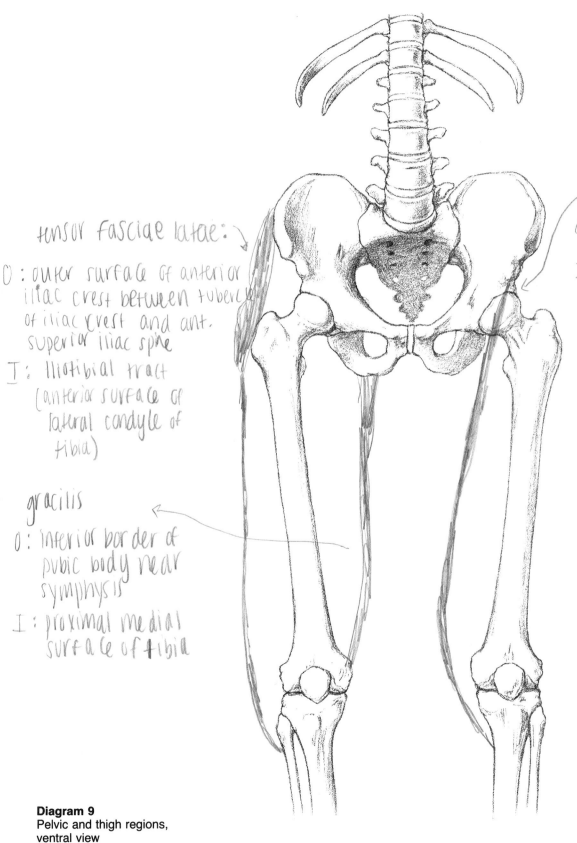

tensor fasciae latae:

O: outer surface of anterior
iliac crest between tubercle
of iliac crest and ant.
superior iliac spine

I: Iliotibial tract
(anterior surface of
lateral condyle of
tibia)

gracilis

O: inferior border of
pubic body near
symphysis

I: proximal medial
surface of tibia

Sartorius:

O: anterior superior
iliac spine

I: proximal tibia,
medial to tibial
tuberosity

Diagram 9
Pelvic and thigh regions,
ventral view

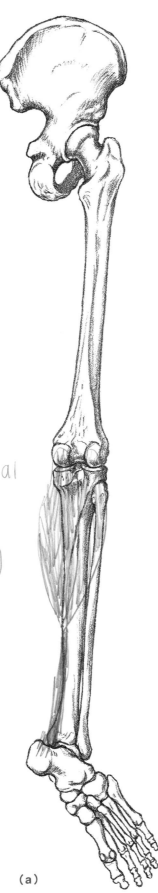

gastrocnemius:
O: lateral and medial
condyle of femur

I: tendo calcaneous
(achilles tendon)

Diagram 10
Hip, thigh, leg, and foot,
showing plantar surface of foot,
dorsolateral view

(a)

(b)

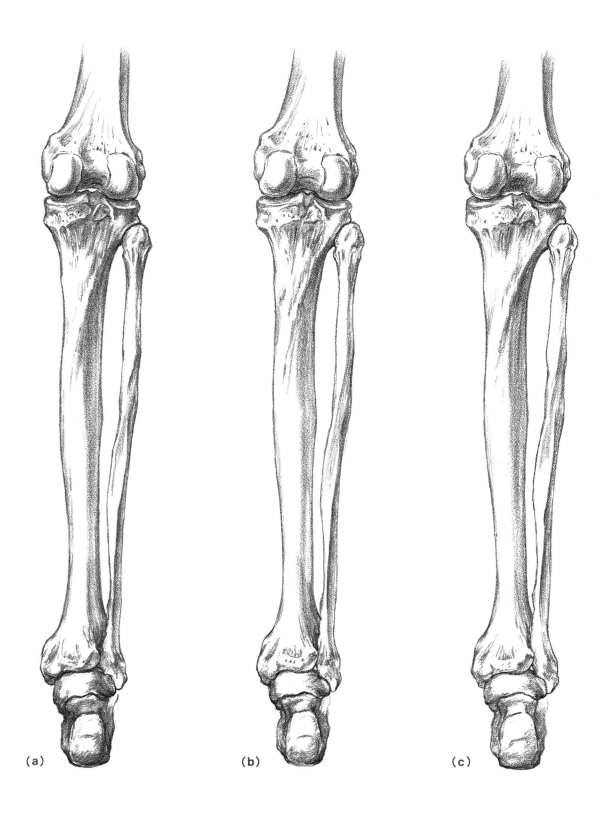

Diagram 11
Leg and foot, dorsal view

(a)　　　　　　(b)　　　　　　(c)

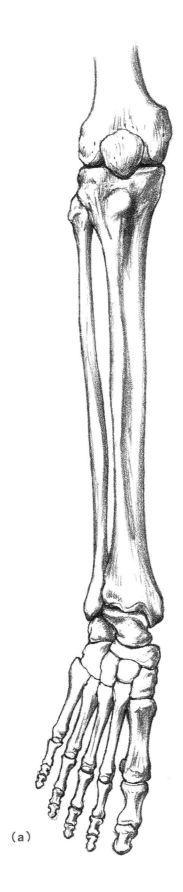
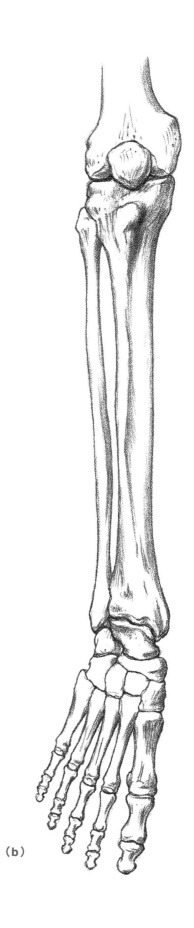

(a)

(b)

Diagram 12
Leg and foot, showing dorsum
of foot, ventral view

4 RESPIRATORY AND DIGESTIVE SYSTEMS

HEAD AND CERVICAL REGIONS

The study of the respiratory and digestive systems should, perhaps, properly begin with the nasal, oral, and pharyngeal areas, which are closely related in embryonic development and remain closely related in the adult. However, at present the only structures which you will study in the head region are the salivary glands. You will observe the other structures of the head region at a later time, when you make a sagittal section of the head (see Chapter 7).

Salivary Glands
(Fig. 3-3, 8-12)

Locate the major salivary glands on either the right or the left side:

Parotid Located ventral to the ear. Its duct crosses the masseter muscle before opening into the vestibule of the oral region.

Submandibular Located ventral to the parotid gland. Its duct passes deep to the digastric muscle and opens into the floor of the oral cavity. This gland was formerly called the **submaxillary.**

Sublingual This gland is quite small. It is deep to the submandibular gland, and appears to be a part of the submandibular gland that extends ventrad. The duct courses with the submandibular duct and opens with it into the floor of the oral cavity.

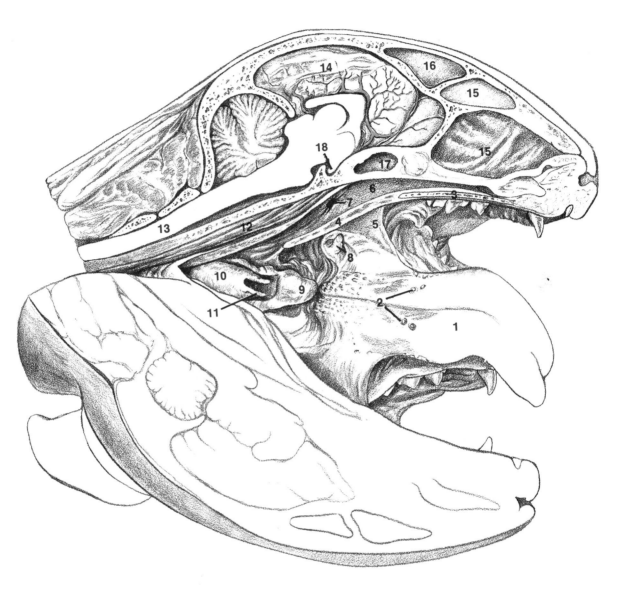

Figure 4-1
Sagittal section of the head

1 Tongue
2 Circumvallate papillae
3 Hard palate
4 Soft palate
5 Wall of oral pharynx
6 Wall of nasal pharynx
7 Opening of internal auditory
  tube
8 Palatine tonsil
9 Epiglottis

10 Ventral wall of esophagus
   at opening
11 Laryngeal cartilage
12 Pharyngeal muscles
13 Spinal cord
14 Brain
15 Nasal septum
16 Frontal sinus
17 Sphenoid sinus
18 Hypophysis

Sagittal Section of the Head
(Fig. 4-1)

This study will be postponed until a sagittal section of the head is made, but the observations that you should make are indicated below.

External nares, or **nostrils** The openings into the nasal cavity.

Nasal cavity Located above the hard palate. Note the septum that divides the cavity into right and left chambers, each with an external naris.

Internal nares, or **choanae** Openings at the caudal end of the nasal septum, between the nasal cavity and the next portion of the respiratory pathway, the nasal pharynx.

Oral opening, or **mouth** The opening into the vestibule of the oral region.

Oral vestibule The space immediately internal to the lips and cheeks, but external to the jaws and teeth.

Tongue The muscular structure protruding from the floor of the oral cavity and the oral pharynx. Note the papillae, particularly the circumvallate papillae on the posterior portion.

Palate The partition, composed of hard and soft portions, between the most cranial parts of the respiratory and digestive pathways. It forms the roof of the oral cavity and oral pharynx, and the floor of the nasal cavity and nasal pharynx.

Oral cavity The portion of the digestive pathway below the hard palate.

Oral pharynx, or **oropharynx** The portion of the digestive pathway below the soft palate.

Nasal pharynx, or **nasopharynx** The portion of the respiratory pathway above the soft palate.

Laryngopharynx Caudal to the oral pharynx and nasal pharynx. The respiratory and digestive pathways cross in the laryngopharynx. Note the openings of the larynx and esophagus, the latter being dorsal.

Openings of the internal auditory tubes (Eustachian tubes) One on each side, in the lateral wall of the nasal pharynx.

Palatine, or **faucial tonsils** Lymphoid organs, one on each side, in the lateral wall of the oral pharynx, lying in the tonsillar sinus, or crypt. If your palatine tonsils have not been removed, you can look in a mirror, open your mouth wide, and note a tonsil on each side in the oral pharynx. The tonsil should appear between two lateral folds of tissue. The cranial fold, which marks the transition between oral cavity and oral pharynx, is called **palatoglossal arch,** and the caudal fold is the **palatopharyngeal arch.** These folds are often called the "pillars of the fauces," the **fauces** being the opening between the oral cavity and the oral pharynx. These "pillars" are not easily observable in the preserved cat. Also, when you look in the mirror, note the extension of the caudal border of your soft palate into a pendulous process. This process is called the **uvula,** and is not present in the cat.

Although you will observe only the palatine tonsils, there is a "ring" of lymphoid tissue in the pharynx region. A **pharyngeal tonsil** lies in the roof and posterior wall of the nasal pharynx. It extends laterally and anteriorly on each side, into the pharyngeal recess behind the opening of the internal auditory tube. When the pharyngeal tonsil is quite enlarged, the condition is called **adenoids.** Completing the ring of lymphoid tissue is the **lingual tonsil,** beneath the mucosa of the caudal portion of the tongue.

Epiglottis The mucosa-covered cartilage ventral to the opening of the larynx. Projects upward, dorsal to the root of the tongue.

Vestibule of the larynx Between the dorsolateral extensions (aryepiglottic folds) of the epiglottis and the ventricular folds.

Ventricular, or **vestibular folds** (false vocal cords) Mucosal folds (one on each side) between the epiglottis ventrally and small cartilages of the larynx dorsally.

Ventricle of the larynx The recess between a ventricular fold and a vocal fold, on each side.

Vocal folds (true vocal cords) One on each side, caudal to the ventricular fold and the ventricle.

Glottis Term often applied to the opening into the larynx and/or the opening between the vocal folds, but it includes the vocal folds.

Note the position of the brain in relation to the respiratory and digestive pathways.

Larynx
(Figs. 3-3, 4-1, 4-2, 4-3)

If the larynx is to be studied in detail, you will do so after the sagittal section of the head has been made.

Certain features of the larynx were listed in the preceding section: the epiglottis, vestibule, false vocal cords, ventricle, true vocal cords, and glottis. Note the small projections in the dorsal wall of the vestibule. These are projections of the **arytenoid cartilages.**

Expose the **thyroid** and **cricoid cartilages** ventrally (using a probe and/or scalpel as needed). The cricoid cartilage is just craniad of the trachea, and the thyroid cartilage is immediately craniad of the cricoid. There are other small cartilages, but they do not have the importance of the ones named.

The hyoid bone is just craniad of the thyroid cartilage.

Note the small external **cricothyroid** muscles (Fig. 3-3) that assist in operating the laryngeal cartilages.

On one side, slightly laterad of the midline, cut the thyroid cartilage on its longitudinal axis—*just barely* through the cartilage so that the underlying muscle (the thyroarytenoid) is left intact. Carefully separate the section of thyroid cartilage from the muscle, and disarticulate it from the cricoid cartilage. Identify the following muscles that attach to the small triangular arytenoid cartilage:

Thyroarytenoid(eus) A thin, broad muscle lying lateral to the vocal folds.

Lateral cricoarytenoid(eus) A small muscle inferior to the thyroarytenoid.

Posterior cricoarytenoid(eus) From the dorsal surface of the cricoid cartilage, the fibers of this muscle run cranially and laterally to the arytenoid cartilage.

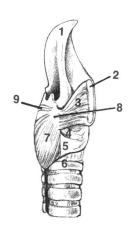

Figure 4-2
Larynx with thyroid cartilage removed (right side), lateral view

1 Epiglottis
2 Thyroid cartilage
3 Thyroarytenoid muscle
4 Lateral cricoarytenoid muscle
5 Cricoid cartilage
6 Cartilage ring of trachea
7 Posterior cricoarytenoid muscle
8 Location of arytenoid cartilage
9 Transverse arytenoid muscle

Transverse arytenoid(eus) A tiny unpaired muscle between the arytenoid cartilages. It is comparable to the transverse portion of the human arytenoideus.

Consult your textbook for the actions of these muscles.

Trachea and Esophagus

(Fig. 4-3)

Note the trachea leading caudally from the larynx, and the two-lobed thyroid gland at the laryngotracheal junction. The cartilage rings in the wall of the trachea are **C**-shaped, with the opening of the **C** being dorsal. Find the esophagus, a muscular tube immediately dorsal to the trachea, and slightly to the left. The trachea and esophagus will be considered again with the thoracic region.

THORACIC REGION

(Fig. 4-3)

The thorax is that portion of the trunk of the body that is craniad of the diaphragm.

Body Wall

Make an incision in the ventral thoracic body wall, about one-fourth to one-half inch to the left of the midline. This should pass through the costal cartilages, which can be cut with a scalpel. Extend the incision cranially to the apex of the thorax (at the base of the neck) and caudally to the cranial surface of the diaphragm.

The muscle just internal to the costal cartilages is the transversus thoracis. The thoracic body wall, from outermost layer to innermost layer, is composed of the following: integument, subcutaneous fascia, muscles and bony structures and their binding tissues, and **fascia endothoracica** (a thin layer of connective tissue). Internal to the fascia endothoracica is the parietal pleura.

Pleura and Pleural Cavities

The pleura is a serous membrane that forms a "pleural sac" on each side of the thorax. The space enclosed by a pleural sac is called a **pleural cavity.** The pleural cavities, like other closed body cavities that you will study, are derivatives of the embryonic coelom. The pleura lining the internal surface of the thoracic body wall is called the **parietal pleura,** and that covering the lungs, which push laterad into the medial wall of their respective sacs, is called the **visceral,** or **pulmonary pleura.** The parietal pleura is continuous at the middorsal and midventral lines with the **mediastinal pleura** (the **mediastinum** being the interval between the pleural sacs), which in turn is continuous with the visceral pleura. The mediastinal and visceral pleurae thus form the medial wall of a pleural sac.

In the living animal, the pleural cavities contain a little serous fluid secreted by the pleural membranes. Note the smooth surface of the membranes. The fluid keeps the membrane surfaces moist and slippery, which allows them to move against each other without friction.

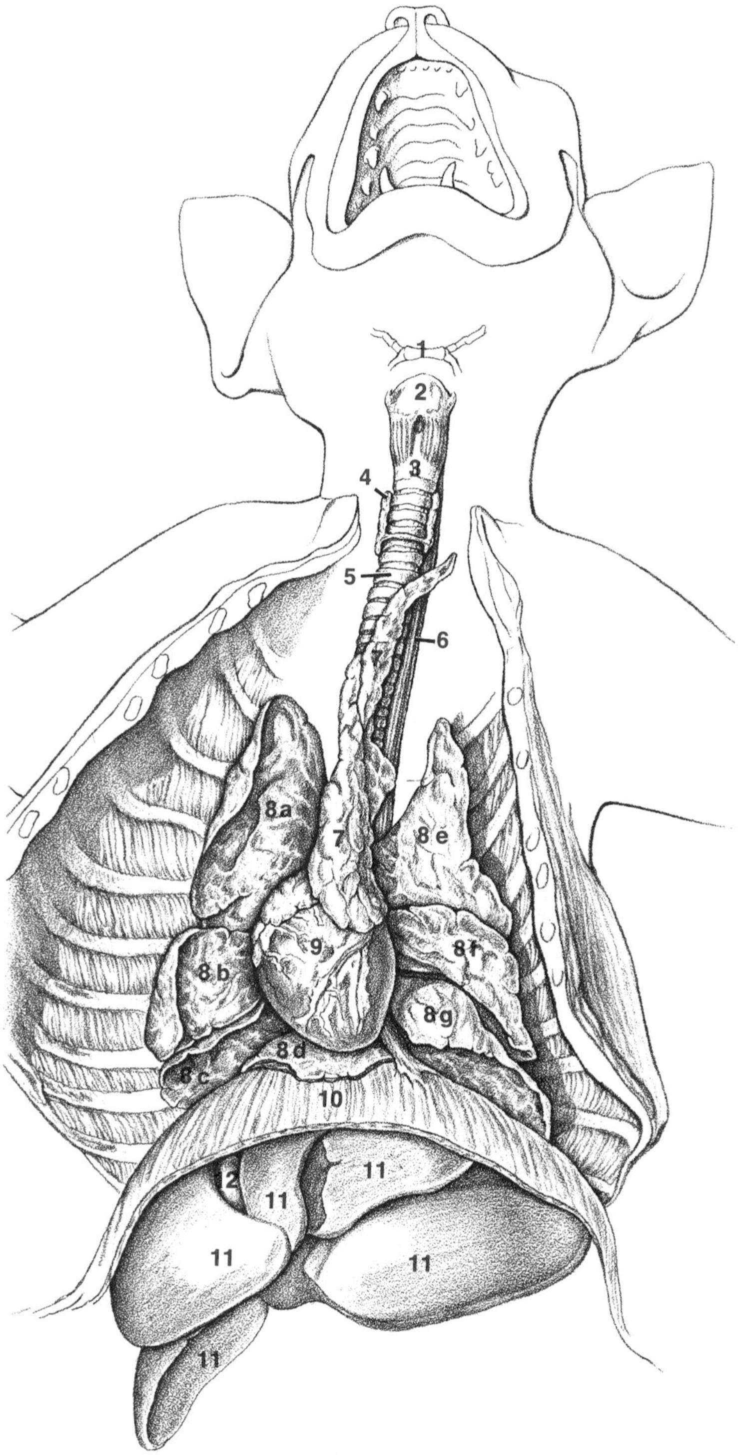

Figure 4-3
Thoracic viscera

  1 Hyoid bone
  2 Thyroid cartilage
  3 Cricoid cartilage
  4 Thyroid gland
  5 Trachea
  6 Esophagus
  7 Thymus gland
  8 Lungs
     a Right, anterior lobe
     b Right, middle lobe
     c Right, posterior lobe
     d Right, mediastinal lobe
     e Left, anterior lobe
     f Left, middle lobe
     g Left, posterior lobe
  9 Heart
 10 Diaphragm
 11 Liver
 12 Gall bladder

Thoracic Cavity

The thoracic cavity is the space, potential or actual, internal to the enthoracic fascia. At the center of the thoracic cavity and extending from the dorsal to the ventral body wall, and from the apex of the thorax to the diaphragm, is the mediastinum (the interval between pleural sacs). The mediastinum contains the major thoracic viscera except for the lungs, and these project laterad from the mediastinum.

Pull the rib cage aside and note the continuity of the mediastinal pleura dorsally and ventrally with the parietal pleura, as well as caudally with the pleura over the diaphragm (this part of the parietal pleura is called **diaphragmatic pleura**).

Observe that the two layers of mediastinal pleura meet ventrally to form a **mediastinal septum.**

Lungs

Detach the rib cage from the diaphragm on each side by cutting along the cranial surface of the diaphragm, just under the caudal border of the rib cage. Extend these incisions far enough laterad and dorsad to expose the lungs. Note the position of the heart within the mediastinum. Using bone shears, cut the ribs dorsally, a short distance laterad of the vertebral column on each side, and pull the rib cage back. This allows a better observation of the thoracic viscera.

Note the thin smooth covering of visceral pleura over the lobes of the lungs and its continuity with the pleura of the mediastinum. Note the soft rubbery texture of the lung tissue. In the living animal, the lungs are extremely soft and spongy and are inflated, so that they essentially obliterate the space in their respective pleural cavities.

Observe that the cat has four lung lobes on the right and three on the left. The human has three lobes on the right and two on the left. In the cat, the most caudal lobe on the right (the mediastinal lobe) passes dorsad and to the left of the postcaval vein, and into a pocket between the mediastinal septum and the caval fold (a fold of pleura extending ventrad from the postcaval vein.

Trachea

Using a probe and/or scalpel as needed, extend the ventral incision cranially through the muscles and fascia of the cervical region to the larynx.

Observe the small two-lobed **thyroid gland.** One lobe lies on each side of the trachea at the caudal border of the larynx. The lobes are connected across the midline, ventral to the trachea, by a thin band of glandular tissue called the **isthmus.** Because the isthmus in the cat is very small, it is usually overlooked and destroyed. The glandular tissue that is ventral to the trachea in its caudal half, and extends to the heart, is the **thymus gland.** The extent of the thymus gland will vary, depending on the age of the specimen. If the cat is a fairly old one, there will be very little thymus tissue present. The thymus gland is concerned with development of the immune response system. Since this development occurs early in life, the thymus regresses as the animal (including the human) grows older. Whatever the size of the thymus gland, it should be left intact for now.

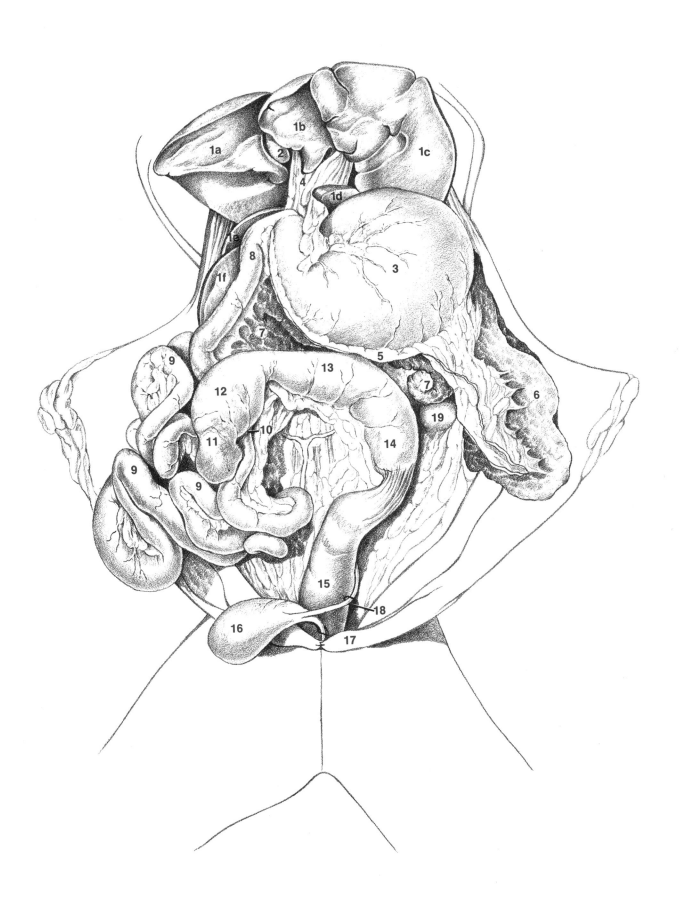

Dorsal to the arch of the aorta, the trachea divides into right and left major bronchi (which also contain **C**-shaped cartilage rings), but you will not be able to observe the division at this time.

Esophagus

The esophagus is a muscular tube dorsal to the trachea and extending a little to the left of it. Probe along the cranial half of the trachea on the left, and partially separate the trachea and esophagus. Do not disturb the thymus gland and blood vessels in the area. The esophagus can also be seen farther caudad on the left, just craniad of the diaphragm, where it lies dorsal to the heart and ventral to the aorta.

ABDOMINAL REGION

(Fig. 4-4)

The abdominal region is that part of the trunk of the body caudal to the diaphragm. The **abdomen proper** is craniad of the pelvic inlet, and the portion caudad of the inlet is called the **lesser pelvis.**

Body Wall

Caudad of the diaphragm, make a cranial to caudal incision just to the left of the ventral midline, through the rectus abdominis and the aponeuroses of the ventrolateral abdominal muscles, and through the **fascia transversalis** (a thin layer of connective tissue comparable to the fascia endothoracica) and **parietal peritoneum** (a serous membrane comparable to the parietal pleura) into the **peritoneal cavity.** From this ventral incision, make another incision on each side along the caudal border of the diaphragm to the dorsal body wall. Pull the resulting flaps of body wall laterad to expose the abdominal viscera that project into the peritoneal cavity.

Diaphragm

This is a dome-shaped, musculotendinous partition between the thorax and abdomen. Note the **central tendon,** the muscular portion, and the structures passing through the diaphragm. The opening through which the aorta passes, between **right** and **left crura** (singular: **crus**), is called the **aortic hiatus.** The opening for passage of the esophagus is called the **esophageal hiatus.** The large vein passing through the **vena caval foramen** is the inferior vena cava.

The diaphragm has three parts: a **sternal** part with origin from the xiphoid process of the sternum; a **costal** part with origin from the cartilages and adjacent parts of the lower six ribs on each side; and a **crural** part arising from the cranial three or four lumbar vertebrae. The fibers insert into the central tendon, which is much less prominent in the cat than in the human. The nerve supply is furnished by ventral rami of cervical spinal nerves by way of the phrenic nerve.

The diaphragm was thought to be a muscle of inspiration only, but recent investigations with dogs found that it acts as two muscles. Contraction of the costal part of the diaphragm increases the dimensions of the lower rib cage (therefore qualifying it as a muscle of inspiration) and contraction of

Figure 4-4
Abdominal viscera

1 Liver
  a Right median lobe
  b Left median lobe
  c Left lateral lobe
  d Caudate lobe
  e Right lateral lobe, anterior (cranial) part
  f Right lateral lobe, posterior (caudal) part
2 Gall bladder
3 Stomach
4 Lesser omentum
5 Cut edge of omental bursa of greater omentum
6 Spleen
7 Pancreas
8 Duodenum
9 Jejunum and ileum
10 Ileocecal junction
11 Cecum
12 Ascending colon
13 Transverse colon
14 Descending colon
15 Rectum
16 Urinary bladder
17 Urethra
18 Ureter
19 Kidney (retroperitoneal)

the crural part decreases the dimensions of the lower rib cage as long as abdominal pressure against the diaphragm is not increased (thus giving it an expiratory function).

Peritoneum and Peritoneal Cavity

The peritoneal cavity is the closed body cavity, derived from the embryonic coelom, into which the abdominal viscera project. This cavity extends into the pelvic region. The part of the membrane lining the inner surface of the abdominal wall is called **parietal peritoneum,** and that covering the viscera is called **visceral peritoneum.**

Most of the major abdominal viscera have peritoneal suspensions, or supports, which are described elsewhere in this chpater.

Note that the continuity of the parietal peritoneum, peritoneal supports, and the visceral peritoneum creates a "peritoneal sac" similar to a pleural sac, and the abdominal viscera have the same relationship to the peritoneal sac and its enclosed cavity as a lung has to its pleural sac and pleural cavity.

Like the pleural cavities, the peritoneal cavity contains a little tissue fluid produced by the serous peritoneal membranes, so that as the viscera move against each other in the living animal there is no friction.

Abdominal Cavity

The abdominal cavity is potential space internal to the fascia transversalis, and it contains all of the major abdominal viscera. Note the relationships of the viscera when you first open the peritoneal cavity. Note the apron, with fat deposits, covering the intestinal portion of the digestive viscera.

Stomach

Lift the lobes of the liver (Figs. 4-3, 4-4) and note the position of the stomach. Observe the esophagus passing through the diaphragm to join the stomach (gastroesophageal junction) at its **cardiac** end. Note the **pyloric** portion of the stomach and the constriction that marks the union of the stomach with the intestine (gastroduodenal junction), and the location of the pyloric valve. The **pyloric sphincter muscle,** which produces the valve action, is a thickening of the circular muscle layer at the caudal end of the stomach. Note the **greater** and **lesser curvatures** of the stomach, the greater on the left, and the lesser on the right.

Spleen

The spleen, which is a hemolymph organ and not a part of the digestive system, will be found on the left along the greater curvature side of the stomach, to which it is attached by the **gastrosplenic ligament.**

Intestine

The intestinal portion of the digestive tube is divided into a **small intestine** and a **large intestine.** The small intestine consists of the **duodenum** (the cranial and shortest portion), **jejunum,** and **ileum.** The ileum joins the large intestine at the junction of the cecum and the colon (ileocolic junction). The large intestine consists of the **cecum, colon** (which has ascending, transverse, and descending portions), **rectum,** and **anal canal.** In the human there is also an S-shaped portion of the colon, between the descending colon and rectum, which is called the *sigmoid colon.* The digestive tube, which begins cranially with the oral opening, or mouth, ends caudally with the anal opening or **anus.**

An outer longitudinal layer of smooth muscle, which is present in the digestive tube, is divided into three bands in the large intestine of the human, and these bands are so arranged as to cause the large intestine to have a sacculated appearance. This muscle layer is not divided in the cat, so the sacculations do not occur.

Peritoneal supports

Lesser omentum This is a double layer of peritoneum extending from the lesser curvature of the stomach and from the duodenum to the liver. The portion between the stomach and the liver is the **hepatogastric ligament;** that between the duodenum and the liver is the **hepatoduodenal ligament.** The lesser omentum contains blood and lymphatic vessels, nerve fibers, and bile ducts between its layers. The **common bile duct** is enclosed in the free edge and can be observed joining the duodenum.

Dorsal to the free border of the lesser omentum, an opening, the **epiploic foramen,** or **foramen of Winslow,** leads into the **lesser peritoneal cavity,** which is dorsal to the lesser omentum and stomach and extends into the omental bursa (which is described in the following paragraph). The portion of the peritoneal cavity that the abdominal viscera project into is called the **greater peritoneal cavity.**

Greater omentum Note the double layer of peritoneum that attaches the stomach, by its greater curvature, to the dorsal body wall. This is the greater omentum, which encloses the spleen and part of the pancreas between its layers. The apron-like, double-layered sac extending caudad, ventral to the intestine, is called the **omental bursa** or **lesser peritoneal sac,** and it is a part of the greater omentum, as is the gastrosplenic ligament between the stomach and spleen.

Lift the omental bursa and spread it out. Note the fat deposits. Carefully separate the dorsal and ventral walls, and note the space (part of the lesser peritoneal cavity) within.

Mesentery The mesentery suspends the small and large intestine from the dorsal body wall. The part that suspends the small intestine is called the **mesentery proper.** Note the great difference in length between the body wall attachment and the small intestine. The part of the mesentery suspending the colon is called the **mesocolon** and that suspending the rectum is called the **mesorectum.** Note the lymph nodes in the mesentery, and the blood and lymphatic vessels coursing between the layers.

Liver, Gall Bladder, and Bile Ducts
(Figs. 4-3, 4-4, 4-5)

Note the lobes and ligaments of the liver, which is a very large and highly vascular organ. The **falciform ligament** is a fold of peritoneum, between the liver and the diaphragm and ventral body wall, that essentially divides

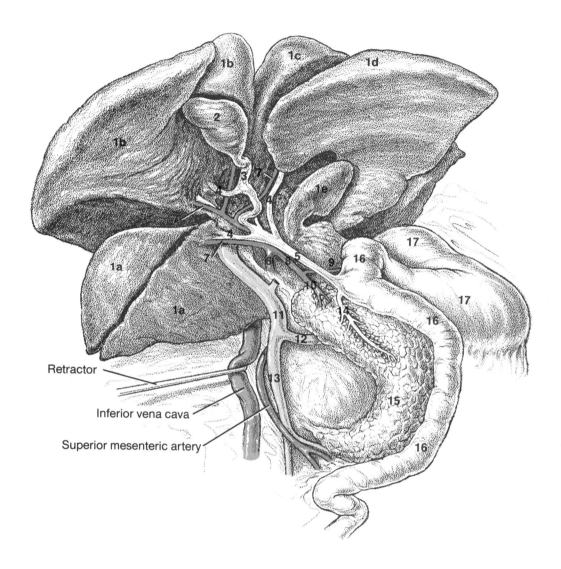

Retractor

Inferior vena cava

Superior mesenteric artery

Figure 4-5
Liver, gall bladder, and associated
vessels. (Lymphatic vessels and
nerve fibers will be present,
although not shown here.)

1 Lobes of liver
  a Right lateral
  b Right median
  c Left median
  d Left lateral
  e Caudate
2 Gall bladder
3 Cystic duct
4 Hepatic ducts
5 Common bile duct
6 Hepatic artery
7 Branches of hepatic artery

8 Gastroduodenal artery
9 Right gastroepiploic artery
10 Superior pancreaticoduodenal
    artery
11 Hepatic portal vein
12 Gastrosplenic vein
13 Superior mesenteric vein
14 Pancreatic duct (duct of
    Wirsung)
15 Head of pancreas
16 Duodenum
17 Stomach

the liver into right and left halves. The **round ligament,** which represents the vestige of an embryonic blood vessel, is a fibrous strand within the free border of the falciform ligament. Note the short suspensory ligaments (**coronary ligaments**) attaching the liver to the diaphragm.

You will find the **gall bladder** (Fig. 4-5) partially concealed in a depression in the right median lobe of the liver. The ducts of the gall bladder and the liver are in the lesser omentum, and great care must be taken in dissecting for these so that nearby blood vessels are not destroyed. The **hepatic ducts** from the liver lobes and the **cystic ducts** from the gall bladder lead into the **common bile duct,** which empties into the duodenum. Within the wall of the duodenum, the common bile duct is joined by the pancreatic duct, and the two have a common opening into the duodenum.

Pancreas

Reflect the omental bursa craniad, and find the pancreas. You can cut the bursa near its attachment to the stomach and spleen, but leave enough attached so that the blood vessels running near the spleen and the greater curvature of the stomach will not be destroyed.

The **head** and **neck** of the pancreas lie in the curve of the duodenum, within the mesentery proper; the **body** extends to the left, between layers of the greater omentum, dorsal to the stomach, and the **tail** extends as far to the left as the spleen. (In the human processes of development cause the pancreas to assume a retroperitoneal position.) The **pancreatic duct** (duct of Wirsung) joins the common bile duct within the wall of the duodenum. Take care not to destroy the blood vessels as you try to locate the duct. There may be an accessory pancreatic duct (duct of Santorini) opening independently into the duodenum. This duct would be from the body and tail of the pancreas.

5 UROGENITAL SYSTEM

The urogenital (or urinogenital) system includes both the urinary (excretory) and reproductive systems. These systems are closely associated in embryonic development and maintain a close relationship in the adult.

URINARY SYSTEM
(Figs. 5-1, 5-2, 5-4)

Locate the **kidneys** lying against the dorsal body wall caudad of the diaphragm. They are not suspended by peritoneum, but are retroperitoneal. In the cat the right kidney lies slightly more craniad than the left one; the positions are reversed in the human. In the cat an adrenal (or suprarenal) gland is located at the cranial end of each kidney, but it is separate and lies slightly mediad of it; in the human the gland actually "caps" the tip of the kidney.

Find the **ureters,** which lead to the **urinary bladder,** coursing along the ventral surface of the psoas muscles. The ureters are also retroperitoneal, and you must use a probe to break and pull aside the peritoneum in order to find them. The urinary bladder is supported, close to the ventral body wall, by a ventral suspensory ligament and by lateral ligaments. The lateral ligaments usually contain fat. The duct leading caudally from the bladder is the **urethra.** The female urethra opens into a vestibule, serving both the urinary and the reproductive systems, which in turn opens to the exterior.

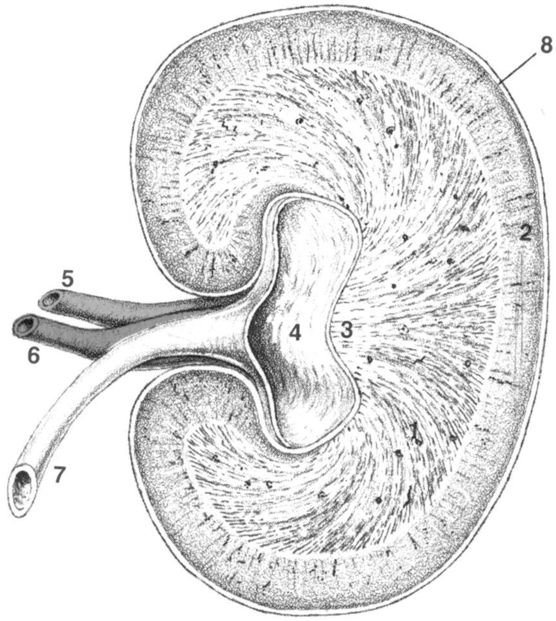

Figure 5-1
Kidney, median longitudinal section

1 Medulla
2 Cortex
3 Renal papilla
4 Renal pelvis
5 Renal artery
6 Renal vein
7 Ureter
8 Renal capsule

The urinary bladder of the male cat (but not of the human) has a "neck," which extends from the bladder to the prostate gland, where it becomes continuous with the prostatic portion of the urethra. You will examine the urethrae in more detail when you study the reproductive system.

Remove the peritoneum from the ventral surface of a kidney. Note the connective tissue (perirenal fascia) around the kidney. This fascia is very thin over the ventral surface of the kidney. Laterally and dorsally it is thicker and may contain a quantity of fat. On the medial side of the kidney note the **hilus,** the point though which the blood vessels and the ureter enter, or leave, the kidney. Make a longitudinal section of the kidney, dividing it into ventral and dorsal halves. Observe the **cortex,** the **medulla,** and the **pyramid,** which has been formed by the coalescence of collecting tubules. The pyramid projects into the **renal pelvis,** which is the expanded portion of the ureter within the **renal sinus** (the space within the hilus). There are several pyramids in the human kidney. The thin layer of connective tissue immediately around the kidney is the **capsule** of the kidney. Peel this away from one section but leave it attached at the hilus.

FEMALE REPRODUCTIVE SYSTEM
(Figs. 5-2, 5-3)

*Locate the **uterus** dorsal to the urinary bladder and urethra. Note the **body** of the uterus and the **cornua,** or horns, extending laterad and craniad from the body. (The human has no uterine horns.) On each side the horn is continuous with the small **uterine tube,** or **oviduct,** which curves around the **ovary.** The uterine tube opens into the peritoneal cavity through its ostium, so the peritoneal cavity is not completely closed in the female.

*Your specimen may have had a neutering operation, a procedure in which the uterus and ovaries are removed. If so, you must observe these structures on another student's cat.

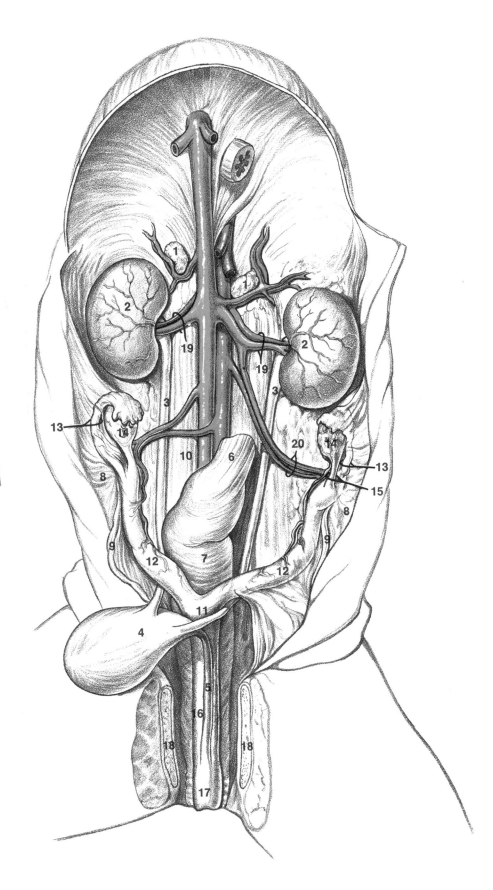

Figure 5-2
Female urogenital system

1 Adrenal gland
2 Kidney
3 Ureter
4 Urinary bladder
5 Urethra
6 Descending colon
7 Rectum
8 Broad ligament
9 Round ligament
10 Psoas muscles
11 Body of uterus
12 Horn of uterus
13 Uterine tube
14 Ovary
15 Ovarian ligament
16 Vagina
17 Urogenital vestibule
18 Pubic bone
19 Renal blood vessels
20 Ovarian blood vessels

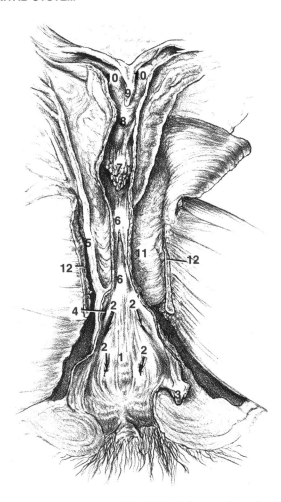

Figure 5-3

Uterus, vagina, and urogenital vestibule of the cat

The ventral wall of the vestibule, vagina, and cervix of the uterus has been slit and pulled laterad. The ventral wall of the uterine body, and of a portion of the cornua, has been removed.

  1 Urogenital vestibule
  2 Openings of vestibular glands
  3 Clitoris
  4 External urethral orifice
  5 Urethra
  6 Vagina
  7 Cervix of the uterus
  8 Body of the uterus
  9 Intrauterine septum
 10 Cornu of the uterus
 11 Levator ani muscle
 12 Cut edge of pubic symphysis

Note the position of the ovary caudad of the kidney. Identify the ligaments: the peritoneal **broad ligament** of the uterus, the **round ligament** of the uterus (the narrow fibrous band enclosed between the layers of the broad ligament), the **mesovarium** (a portion of the broad ligament attached to the ovary and containing vessels to the ovary), and the **ovarian ligament** (a short, thick cord within the broad ligament, extending from ovary to uterus).

Cut through the pubic symphysis and bend the thighs dorsad to loosen the attachments. It may be necessary to cut away some of the pubic bone with bone shears. Separate binding tissues to expose the urethra, the body of the uterus, and the **vagina,** which is dorsal to the urethra. Follow the course of the urethra and vagina caudally to the site at which both open into the **urogenital vestibule,** which in turn opens to the exterior. The urogenital vestibule of the cat is much deeper than that of the human, which is quite shallow. The position of the **cervix,** or neck, of the uterus can sometimes be determined externally: it is a knot of tissue (dorsal to the urethra) that is usually somewhat harder than the body of the uterus just craniad, and the vagina just caudad.

Note the dorsal position of the rectum and anal opening.

Note the flat sheet-like muscle, the **levator ani,** lateral to the rectum (see Fig. 8-18). This muscle has origin on the pubis and ischium and inserts into tissues around the caudal end of the rectum. The muscles from each side insert into a raphe dorsal to the rectum and the two together act as one muscle, furnishing the major part of the **pelvic diaphragm** (the funnel-shaped floor of the pelvis). The obturator internus muscle is just under (lateral to) the levator ani.

The following procedure is optional, depending on the decision of the instructor, and should be done *only after consulting the instructor.*

To observe the internal appearance of the vagina and the cervix of the uterus (Fig. 5-3), make an incision in the ventral wall of the urogenital vestibule a little laterad of the midline. Extend the incision craniad (laterad of the urethral opening) through the wall of the vagina and into the uterine cervix. Pull the cut edges laterad. Note the mucosal folds of the cervix, which can fit together to close the cervical canal. In the urogenital vestibule, the openings of two pairs of glands can be observed. The more caudal ones (**greater vestibular glands,** or Bartholin's glands of the human) are homologs of the bulbourethral glands of the male but are relatively smaller.

If you have kept the incision laterad of the midventral line in the vestibule, you may be able to observe a tiny papilla in the midventral wall. This is the **clitoris,** a homolog of the male penis.

If you extend the incision craniad into the body of the uterus, you will find an internal septum in the cranial portion of the body. To observe the septum, the ventral wall of the body of the uterus and a portion of the ventral wall of the cornua can be cut away, as shown in Figure 5-3.

Consult your textbook for descriptions and illustrations of the human uterus and the external genitalia.

MALE REPRODUCTIVE SYSTEM
(Figs. 5-4, 5-5)

*Caudad of the pelvic region, locate the **penis,** which contains a part of the urethra, and the **scrotum,** an internally divided integumentary sac which encloses the **testes.** Find the two **spermatic cords,** each immediately laterad of the ventral midline (at the pubic symphysis). Each cord contains nerves, and blood and lymphatic vessels, as well as a sperm duct, the **ductus** or **vas deferens.**

Using the probe to loosen attachments a little, and with judicious use of scissors, follow a spermatic cord caudad to the scrotum, snipping connective tissues only as necessary. Make an incision in the ventral wall of the scrotal sac in order to expose the testis. To observe the testis and the **epididymis,** you will have to slit and reflect the covering of these structures.

The testis and epididymis have several covering layers of tissue, which are extensions from the abdominal body wall. These layers also help to form and cover the spermatic cord. The layers are described below, from the innermost to the outermost one.

*If your specimen has had a neutering operation, the testes will have been removed. After such an operation, the scrotum decreases greatly in size and the spermatic cords also atrophy. If this is the case, proceed with the dissection, but observe the missing structures on another student's cat.

Figure 5-4
Male urogenital system

1 Adrenal gland
2 Kidney
3 Ureter
4 Urinary bladder
5 Neck of bladder
6 Prostate gland
7 Pubic bone
8 Urethra
9 Bulbourethral gland
10 Crus of penis
11 Ischiocavernosus muscle
12 Penis
13 Spermatic cord with fascial
    covering
14 Testis and epididymis with
    fascial covering
15 Scrotal integument
16 Epididymis*
17 Testis
18 Ductus deferens*
19 Location of inguinal canal
20 Internal spermatic blood
    vessels
21 Rectum
22 Descending colon
23 Renal blood vessels
24 Psoas muscles
 * The right testis has been
    rotated so that the ductus
    deferens appears ventrally. The
    ductus is on the medial side of
    the testis, and the epididymis is
    dorsal in position.

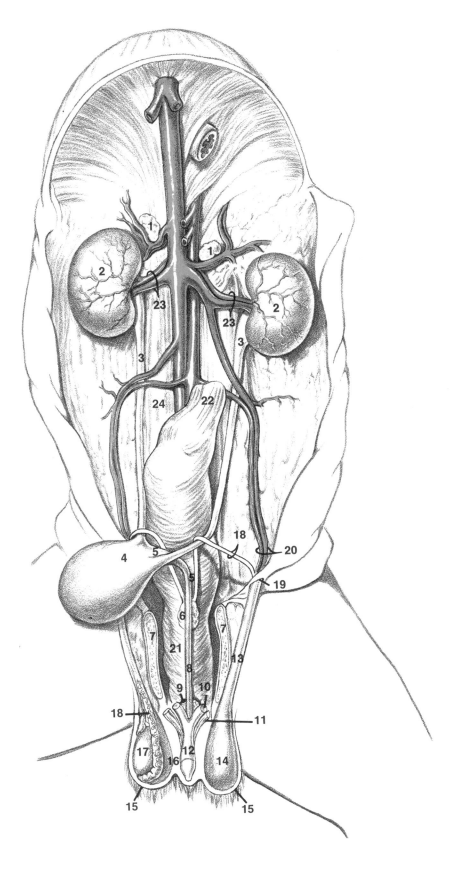

The innermost layers are formed from a double layer of peritoneum that resulted from an "outpouching" (**processus,** or **saccus vaginalis**) at the caudal end of the trunk during development. This peritoneal sac surrounded the testis and associated structures after the testis, which began development higher in the trunk, descended to occupy a position in the pouch. The layer of peritoneum that is closely adherent to the testis and epididymis is the visceral layer and the outer layer of peritoneum is the parietal layer of the **tunica vaginalis** (tunica vaginalis propria). The parietal layer can be slit and peeled back to expose the testis and epididymis, and you will note a continuity of the parietal and visceral layers so that a space is enclosed. The space is from the peritoneal cavity, which extended into the saccus vaginalis. Around the ductus deferens within the spermatic cord, these peritoneal layers normally fuse and atrophy, so that the extension from the peritoneal cavity is sealed off. Thus, the tunica vaginalis is usually not present in the spermatic cord, except for a short extension craniad of the parietal layer, but other extensions from the body wall will be present.

The next three layers blend with the parietal layer of the tunica vaginalis and are not easily identified in the cat as separate entities. You will have slit these layers, along with the parietal tunica vaginalis. First is the **internal spermatic fascia** (tunica vaginalis communis), an extension from the transversalis fascia. Next in order is the **levator scroti layer** (the cremaster layer of the human), which contains some muscle fibers as well as connective tissue, and is an extension from the internal abdominal oblique muscle and its fascia. The third layer is the **external spermatic fascia,** which is an extension from the fascia of the external abdominal oblique muscle.

Outside of the above layers is the superficial fascia, which in the human contains smooth muscle fibers and is called the **dartos tunic.** You cut into this tunic when you opened the scrotum. The scrotum is basically composed of integument and the underlying superficial fascia. The superficial fascia forms an internal septum within the scrotum, so that each testis lies in an individual compartment.

Note the epididymis, which forms a band curving around the testis from the lateral to the dorsal side (Fig. 5-5). The band is composed of a head, body, and tail, with the head being the most lateral in position. You will not see the **efferent ductules** from the testis but they unite, within the head of the epididymis, with the highly convoluted **duct of the epididymis.** The duct remains convoluted as it continues through the body and tail of the epididymis, where it emerges to continue as the ductus deferens. The latter is also convoluted for some distance as it begins its ascent from the testis into the spermatic cord.

Follow the ductus deferens and other structures of the spermatic cord craniad to the **external inguinal ring.** Find the ductus deferens and internal spermatic blood vessels passing through the **internal inguinal ring.** The short channel between the inguinal rings is called the **inguinal canal.** Follow the ductus deferens as it passes ventral to the ureter, then mediad to a position dorsal to the urinary bladder and the neck of the bladder. Note that it courses with its counterpart of the opposite side, within a fold of peritoneum, to the **prostate gland.** The duct passes through the prostate tissue to reach the prostatic portion of the urethra. (In the human a **seminal vesicle** joins the ductus deferens just before it enters the prostate, and the

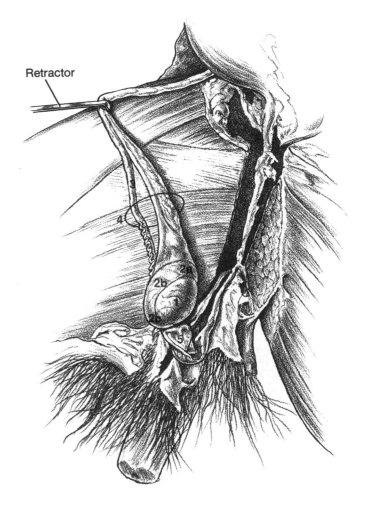

Retractor

Figure 5-5
Right testis and epididymis, dorsal surface

1 Testis
2 Epididymis
  a Head
  b Body
  c Tail
3 Ductus deferens
4 Spermatic cord
5 Parietal layer of tunica vaginalis

portion of the duct passing through the prostatic tissue is called the **ejaculatory duct.**)

Cut through the pubic symphysis and bend the thighs dorsad, to loosen the attachments. It will be necessary to cut away, with bone shears, some of the pubic bone in order to dissect the underlying structures. Carefully break binding tissues to expose the neck of the urinary bladder and the prostate gland (small in the cat). (In the human the prostate gland is immediately caudad of the bladder.) Note the levator ani muscle laterad of the rectum, as described above with the female reproductive system.

The male urethra has three portions: prostatic, membranous, and cavernous. The **prostatic urethra** is the portion passing through the prostate gland, which empties its secretions into the urethra. The **membranous urethra** is the portion between the prostate and the **bulbourethral glands** (Cowper's glands). Locate the bulbourethral gland on one side. To do this, cut the **crus** of the penis (the proximal end of the cavernous body) and a small muscle (the ischiocavernosum) from their attachments on the ischium. The gland will be found dorsal to these structures. Distal to the point at which the ducts of the bulbourethral glands join the urethra is the **cavernous urethra** (also variously called the "spongy urethra" and the "penile urethra"), which courses through the **penis** to open to the exterior through the terminal portion of the penis (the glans penis).

The penis is composed of three cylindrical masses of cavernous tissue with an integumentary covering. The two dorsal and larger cylinders are called the **cavernous bodies.** The smaller ventral cylinder, which contains the cavernous urethra, is called the **spongy body.** The cavernous tissue contains endothelial-lined spaces that fill with blood to cause rigidity of the penis.

Note that the male urethra has a dual function: reproductive and excretory.

Each student should study both female and male specimens.

6 CIRCULATORY SYSTEM

The circulatory system includes the vessels through which the blood flows and an auxiliary system, the lymphatic system. The lymphatic vessels are so thin-walled that it is not practical to do much dissection for them, but major ones can be observed as the blood vessels are studied. The lymphatics will have a yellow or light brown color and are likely to have a beaded appearance, due to the internal semilunar valves throughout the length of the vessels. Many of the smaller lymphatic vessels terminate in lymph nodes (you have previously observed some of these nodes), which act as filtration devices for tissue fluid and also produce lymphocytes and other cells. Other vessels leave the lymph nodes and may terminate in other nodes or join other lymphatic vessels. There are three major lymphatic vessels, and these will be mentioned later in this chapter.

VESSELS CRANIAD OF THE DIAPHRAGM

Heart
(Figs. 4-3, 6-1, 6-2, 6-3)

Note the position of the heart within the mediastinum. Carefully clear away any thymus tissue and fat that may obscure the heart and its attached vessels.

The nerve ventral to the root of the lung on each side is the **phrenic,** which passes to the diaphragm. (The root of the lung consists of the major

(a)   **(b)**

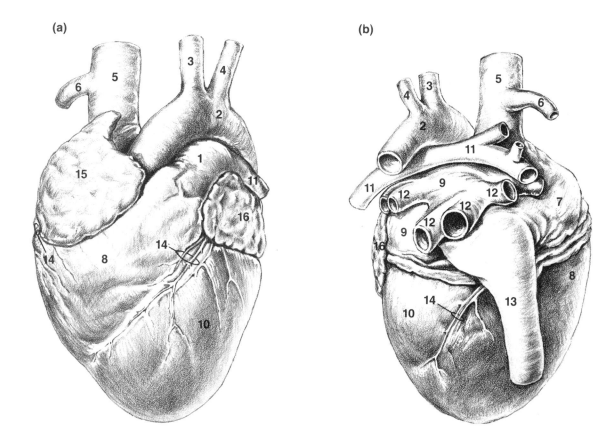

Figure 6-1
The heart and attached vessels
   (a) Ventral view
   (b) Dorsal view

 1 Pulmonary aorta
 2 Arch of aorta
 3 Brachiocephalic artery
 4 Left subclavian artery
 5 Superior vena cava
 6 Azygos vein
 7 Right atrium
 8 Right ventricle
 9 Left atrium
10 Left ventricle
11 Pulmonary arteries
12 Pulmonary veins
13 Inferior vena cava
14 Coronary blood vessels
15 Right auricula
16 Left auricula

bronchus and accompanying vessels and nerve fibers, which are surrounded by mediastinal connective tissue, and have pleura reflected over them as they enter or leave the lung.) On the left side the nerve passing ventral to the arch of the aorta and then dorsal to the root of the lung is the **vagus.** On the right, this nerve can be found along the side of the trachea and then passing dorsal to the root of the lung.

Note the outer wall, the **parietal pericardium,** of the **pericardial sac** that surrounds the heart. Slit this outer wall in a craniocaudal direction and reflect from the heart. Note the attachment of the fibrous outer layer of parietal pericardium onto the blood vessels that join the heart, and the continuity of the inner serous layer with the serous inner wall of the sac. The inner wall is the **visceral pericardium** (epicardium), which invests the heart so closely that it is virtually impossible to separate it from the underlying **myocardium** (heart muscle). The space enclosed by the inner and outer walls of the pericardial sac is the **pericardial cavity,** which is a derivative of the embryonic coelom. As with other derivatives of the coelom, the inner visceral layer can move against the parietal layer without friction.

Note the **right** and **left atria.** Each atrium has a part that is somewhat separated from the main chamber and is called the **auricula,** or **auricular appendage.** Note the **right** and **left ventricles,** with a slight groove between them on the ventral side. **Coronary** blood vessels course in the groove and in sulci between the atria and ventricles. Note the large blood vessels attached to the heart: the **superior** and **inferior venae cavae** joining the

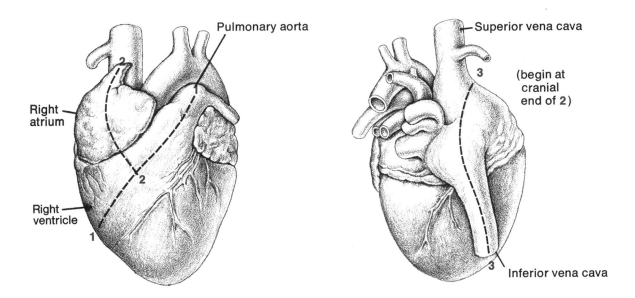

Figure 6-2
Incisions to make when opening the heart

right atrium, the **pulmonary aorta** (or pulmonary trunk) arising from the right ventricle, the **pulmonary veins** joining the left atrium, and the ascending portion of the **aorta** arising from the left ventricle. **Pulmonary arteries** and **veins** can be observed dorsally.

Make incisions in the heart as indicated in the diagrams in Figure 6-2. Incision 1 passes through the ventral wall of the right ventricle into the base of the pulmonary aorta; incision 2 passes through the wall of the right atrium and through the opening of the superior vena cava; incision 3 passes through the opening of the inferior vena cava. Clean out any coagulated blood that may be present in the heart chambers of uninjected specimens. Carefully remove the latex from injected specimens.

Note the difference in thickness and structure of ventricular and atrial walls, and the endothelial lining of the chambers. The endothelium and the underlying connective tissue form the **endocardium,** which is the inner layer of the heart wall. The other layers are the muscle (myocardium) and the visceral pericardium (epicardium). The ridges and projections of muscle on the internal surface of the ventricles are called **columna,** or **trabeculae carneae,** with the tall columns being further designated as **papillary muscles.** In the main part of an atrium, the internal surface is fairly smooth, but the auricular appendage has interlacing ridges of muscle called **musculi pectinati.**

Note the slight depression in the interatrial septum (difficult to observe in specimens that have been injected because of stretching of the septum). This is the **fossa ovalis,** which marks the location of the **foramen ovale** (a fetal opening between atria). The **coronary sinus,** which receives blood from veins that drain the heart wall, opens into the right atrium just medial to the opening of the inferior vena cava.

Note the **atrioventricular valve cusps,** or flaps, guarding the right atrioventricular opening. At their bases the valve cusps are attached to a fibrous

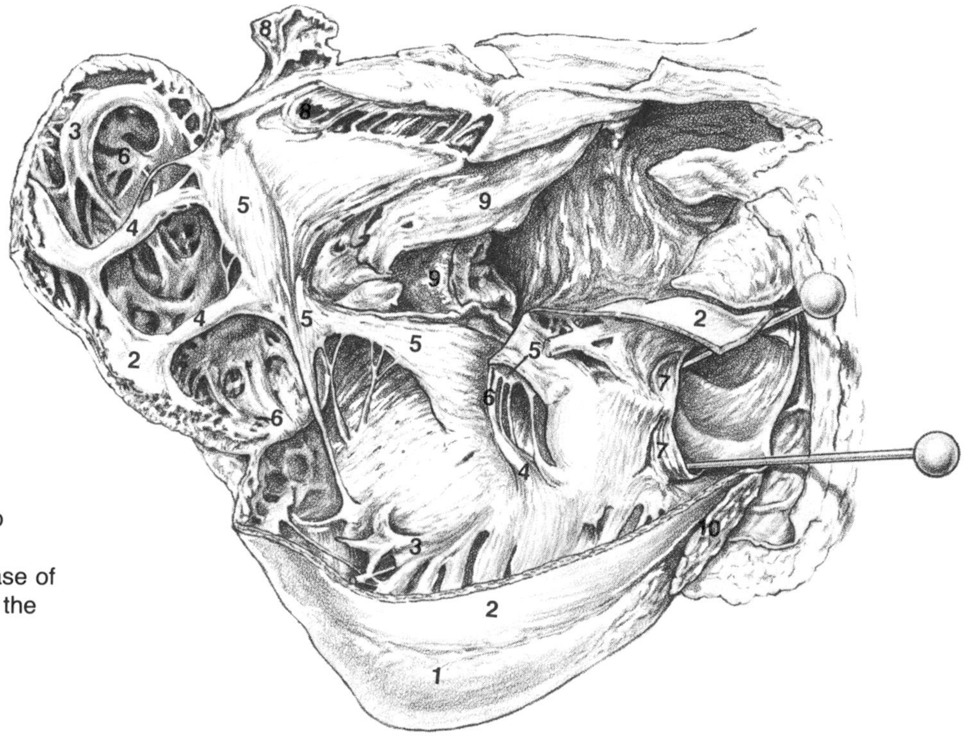

Figure 6-3
Internal structure of the heart (right side)

1 Left ventricle
2 Right ventricle
3 Columna, or trabeculae carneae
4 Papillary muscle
5 Atrioventricular valve cusp
6 Chordae tendineae
7 Semilunar valve flap at base of pulmonary aorta (pin is in the sinus)
8 Auricula of right atrium
9 Right atrium
10 Auricula of left atrium

connective tissue ring between atrium and ventricle. Note the fine **chordae tendineae** attaching the free edges of each cusp to the ventricular muscle. This attachment prevents the cusps from flapping back into the atrium, thus preventing backflow, when the ventricle contracts. The right atrioventricular valve, with three cusps, is called the **tricuspid valve**; the left valve, with two cusps, is called the **bicuspid valve,** or the **mitral valve.**

Note the three **semilunar valve cusps,** or flaps, at the base of the pulmonary aorta. The valve cusp is a fold of endothelium with a little connective tissue between layers of the fold; the **semilunar sinus** (sinus of Valsalva) is the space between the flap and the wall of the vessel. If blood backflows, the sinuses fill with blood and thus close the opening of the vessel.

Because structures on the left are basically the same, except that the wall of the left ventricle is much thicker than that of the right, opening the left side of the heart is optional.

The following directions may vary at the instructor's discretion. *After the study of the nervous system has been completed,* remove the heart from *one specimen at each table,* so that the dorsal side can be observed. Sever vessels in such a way that part of each is left attached to the heart. Cut the aorta at the junction of the ascending portion and the arch, so that the vagus nerve will not be disturbed. When cutting the pulmonary vessels, take care that the vagus and phrenic nerves are not destroyed. Observe the division of the trachea into the right and left major bronchi, as well as the pulmonary vessels between the heart and lungs.

Arteries

(Figs. 6-4, 6-5, 8-16)

Note the general relationships of arteries, veins, and nerves in the mediastinum (see Fig. 6-4). Locate as many of the following arteries as possible, using a probe to separate the vessels from surrounding tissues. Figures 6-4 and 6-5 will aid with identification.

Pulmonary aorta, or **trunk** Carries deoxygenated blood to the lungs via its divisions, the **right** and **left pulmonary arteries.**

Try to find the **ligamentum arteriosum,** a strong band of connective tissue between the pulmonary aorta, at its bifurcation into pulmonary arteries, and the arch of the aorta. This is the remnant of a fetal vessel, the **ductus arteriosus,** that shunted blood to the aorta (the lungs not being functional before birth).

Aorta (systemic aorta) Carries oxygenated blood for all parts of the body. Note ascending, transverse (arch), and descending portions. Note the relationship to the trachea, bronchus, and esophagus. Identify the following branches:

Coronary Paired; from the base of the ascending aorta. The coronary arteries supply blood to the heart wall.

Brachiocephalic (innominate) Unpaired; from the arch. Supplies the head, neck, right pectoral appendage, and ventral body wall on the right.

Left subclavian From the arch. Supplies the left pectoral appendage and left ventral body wall.

Parietal Paired dorsal intercostal, subcostal, and phrenic branches of the descending aorta. They supply body wall structures and the diaphragm.

Visceral The bronchial, esophageal, and pericardial branches of the descending aorta. (Some are paired, some are not.)

In the human there are three branches from the arch of the aorta: the brachiocephalic, the left common carotid, and the left subclavian.

Brachiocephalic This branch of the aorta is described in the preceding list. It has the following branches.

Right subclavian Supplies the right pectoral appendage and ventral body wall.

Right common carotid The carotids are described in the following paragraphs.

Left common carotid (In the cat.)

In the human the branches of the brachiocephalic artery are typically the right common carotid and right subclavian only, with the left common carotid arising directly from the arch of the aorta.

The common carotid artery will be found laterad of the trachea, on each side, in a connective tissue sheath (the carotid sheath), along with the vagus nerve, the cervical portion of the sympathetic trunk, the internal jugular vein, and a lymphatic trunk.

Figure 6-4
Major arteries and veins
of the thorax

1 Heart
2 Lung lobe
3 Aorta
4 Superior vena cava
5 Brachiocephalic artery
6 Sternal vein
7 Left subclavian artery
8 Right subclavian artery
9 Internal mammary arteries
10 Internal mammary artery and
   vein
11 Brachiocephalic vein
12 Subclavian vein
13 External jugular vein
14 Common carotid artery
15 Internal jugular vein
16 Vagus nerve and sympathetic
   trunk
17 Inferior thyroid vein
18 Trachea
19 Phrenic nerve
20 Thyrocervical artery
21 Vagus nerve

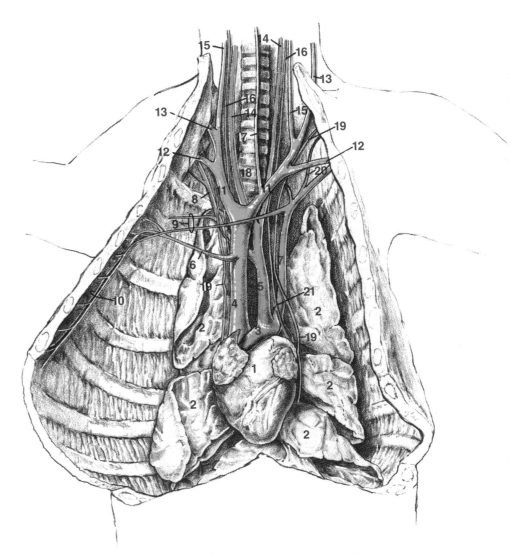

The carotid arteries carry a blood supply for the head region. Near the base of the skull each common carotid divides into an **external carotid**, which supplies primarily the head structures outside the cranial cavity, and an **internal carotid**, which supplies primarily the structures within the cranial cavity. The internal carotid of the cat, which is not entirely analogous to that of the human, is very small and may be degenerate. In the cat the proximal portion of the internal carotid usually becomes a thin ligamentous strand between the common carotid and the medial tip of the tympanic bulla (in the cat, a rounded projection of the mastoid portion of the temporal bone). The distal portion of the internal carotid joins another artery (ascending pharyngeal branch of the occipital artery) within the skull. If the lumen of the internal carotid has remained open, the artery can be located on injected specimens.

Common carotid The branches that may be located on injected cats are the following:

Various small branches to muscles
Superior thyroid

Inferior thyroid Because this is very tiny, it is not usually observed.

Superior laryngeal Supplies the larynx.

Occipital Supplies deep neck muscles and the occipital region.

External and internal carotids Described in the preceding paragraphs.

External carotid Locate the external jugular, anterior facial, posterior facial, and transverse veins (see Fig. 6-6 and p. 143). Do not destroy these superficial veins, but loosen the submandibular and sublingual glands (do not cut the ducts). Loosen the parotid gland, cut the duct, and reflect the gland dorsad. Locate the branches of the external carotid:

Lingual Runs along the ventral border of the digastric muscle, accompanied by the hypoglossal nerve.

Branch to the submandibular and sublingual glands In some specimens this is a branch of the external maxillary.

External maxillary Passes deep to the digastric muscle, gives off a submental artery that passes under the chin, and crosses the lower jaw to terminate as labial arteries that supply tissues around the mouth.

Posterior auricular Passes dorsal to the auricula.

Superficial temporal Ventral to the ear. It branches off deep to the parotid and extends into the temporal region.

Internal maxillary A continuation of the external carotid deep to the masseter muscle. This artery furnishes many branches to the deep tissues of the face, to the orbit, and even sends a branch to the dura mater of the brain.

Subclavian Supplies the head, neck, and ventral thoracic wall, as well as the pectoral appendage. Its branches are the following:

Vertebral Turns dorsad and enters the transverse foramen of the sixth cervical vertebra and passes through the successive transverse foramina en route to the brain. The two vertebral arteries unite on the ventral surface of the medulla oblongata (of the brain) to form the **basilar artery** (Fig. 8-4). The vertebral artery gives branches to the vertebral column and the spinal cord.

Costocervical trunk Supplies costal and cervical regions.

Thyrocervical trunk Continues in the shoulder region as the **transverse scapular** (suprascapular) artery.

Internal mammary (internal thoracic) Supplies the ventral thoracic wall, giving off many ventral intercostal branches, and continues caudad of the diaphragm as the **superior epigastric** artery.

Axillary The continuation of the subclavian artery into the axilla. Its branches are:

Anterior thoracic Supplies the pectoralis muscles.

Long thoracic (lateral thoracic) Supplies the pectorales, the serratus anterior, and the latissimus dorsi.

Subscapular It has two major branches: the **posterior humeral circumflex** to shoulder muscles and dorsal muscles of the arm; and the **thoracodorsal** to the teres major and the latissimus dorsi. Other branches that may be given off in some specimens are the **anterior humeral circumflex** and **deep brachial**, but these are usually branches of the brachial artery. The main trunk of the subscapular continues on to supply the subscapularis and other muscles on the dorsal side of the scapula.

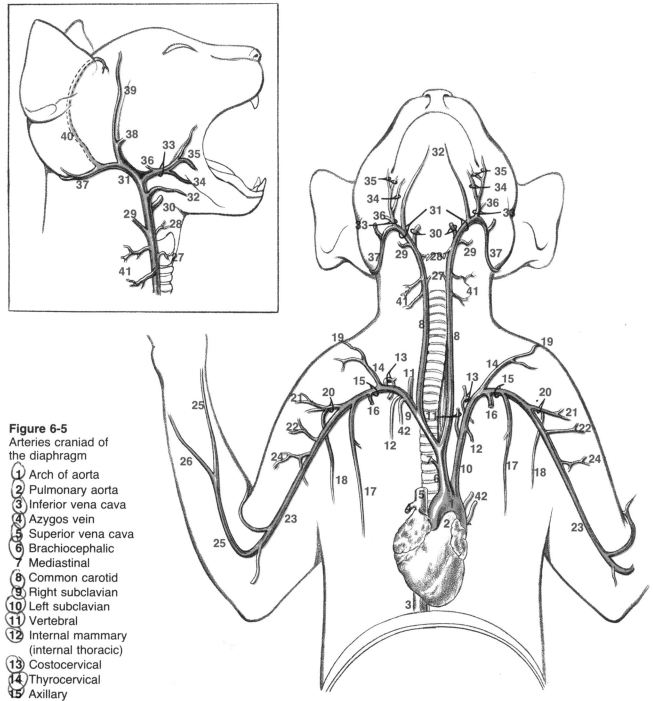

Figure 6-5
Arteries craniad of
the diaphragm

1 Arch of aorta
2 Pulmonary aorta
3 Inferior vena cava
4 Azygos vein
5 Superior vena cava
6 Brachiocephalic
7 Mediastinal
8 Common carotid
9 Right subclavian
10 Left subclavian
11 Vertebral
12 Internal mammary
     (internal thoracic)
13 Costocervical
14 Thyrocervical
15 Axillary
16 Anterior thoracic
17 Long thoracic
18 Thoracodorsal
19 Transverse scapular
20 Subscapular
21 Posterior humeral circumflex
22 Anterior humeral circumflex
23 Brachial
24 Deep brachial
25 Radial

26 Ulnar
27 Superior thyroid
28 Laryngeal
29 Occipital
30 Internal carotid
31 External carotid
32 Lingual
33 External maxillary
34 Submental

35 Superior and inferior labials
36 Branch to submandibular and
     sublingual glands
37 Posterior auricular
38 Internal maxillary
39 Superficial temporal
40 Anterior auricular
41 Branches to muscles
42 Superior intercostal   ("intercostals")

Major arteries of the pectoral extremity

Brachial A continuation of the axillary artery into the arm region. Its branches are:

Anterior humeral circumflex Extends to the biceps brachii and the head of the humerus.

Deep brachial Extends to the triceps brachii, epitrochlearis, and latissimus dorsi.

Various other arteries to muscles

Arteries to the elbow region

Radial The brachial artery continues distal to the elbow as the radial. The radial has various branches, including the small **ulnar** in the cat. In the human the brachial divides into radial and ulnar arteries immediately distal to the bend of the elbow, the ulnar usually being the larger.

In the hand the radial and ulnar arteries give off branches that anastomose to form an arch (arches, in the human), which in turn furnishes branches to the digits.

Veins
(Figs. 6-4, 6-6)

Veins, in general, course parallel to arteries of the same name, but there are some notable exceptions. Refer to Figure 6-4 for general relationships of the major veins, arteries, and nerves in the mediastinum. Locate as many of the following veins as possible, using the descriptions and Figures 6-4 and 6-6 as aids.

Inferior vena cava (postcaval vein) Unpaired vessel joining the right atrium. It returns all blood to the heart from areas caudad of the diaphragm.

Superior vena cava (precaval vein) Unpaired vessel joining the right atrium. It returns all blood to the heart from areas craniad of the diaphragm, except that from the wall of the heart itself.

Azygos Unpaired vessel located on the right side along the vertebral column. It joins the superior vena cava just before the latter joins the right atrium. The azygos collects blood from the dorsal thoracic wall through dorsal intercostal tributaries, and some from the diaphragm. It receives **esophageal veins** from the esophagus and **bronchial veins** from the lungs. The human has accessory azygos veins.

Pulmonary Returns oxygenated blood from the lungs to the left atrium. There may be two or more of these; the human has four.

Brachiocephalic (innominate) The right and left veins join to form the superior vena cava.

Subclavian Joins with the external jugular vein to form the brachiocephalic. The subclavian is a continuation of the axillary vein; therefore it returns blood from the pectoral extremity. In the human the subclavian and internal jugular join to form the brachiocephalic, with the external jugular being a tributary to the subclavian.

External jugular Formed by the union of the **anterior** and **posterior facial veins,** which have tributaries that return blood distributed by the branches of the external carotid artery; therefore the external jugular returns blood primarily from structures of the head region outside the cranial cavity. The two veins communicate near the point of formation by a large **transverse vein,** which passes ventral to the neck. The transverse vein receives tributaries from the tongue and the larynx.

In the cat the external jugular, near its junction with the subclavian vein, is joined on the left by the thoracic duct and on the right by the right lymphatic trunk. In the human these vessels usually join the internal jugular veins at the brachiocephalic junction.

Internal jugular Formed by a union of veins that are primarily from structures within the cranial cavity. It may join either the external jugular or the brachiocephalic in the cat. In the cat the internal jugular is usually quite small and seldom receives an injection of latex, so it is not easily identified. The external jugular vein in the cat is quite large; in the human the internal jugular is the larger.

Vertebral Accompanies the vertebral artery, but does not extend into the cranial cavity, as the artery does. In the cat the vertebral vein joins the costocervical to form a common trunk, which in turn joins the brachiocephalic vein on the left side, and on the right side may join either the brachiocephalic or the superior vena cava. In the human the vertebral vein joins the brachiocephalic directly.

Internal mammary (internal thoracic) In the cat the right and left veins unite to form the **sternal vein,** a tributary to the superior vena cava. In the human the internal mammary vein joins the brachiocephalic. The internal mammary vein receives tributaries corresponding, in course and in name, to the branches of the internal mammary artery.

Transverse scapular (suprascapular) Joins the external jugular. In the cat it is joined by the cephalic vein from the extremity.

Inferior thyroid Joins the left brachiocephalic vein.

Axillary Located in the axilla. This is a continuation of the **brachial vein** from the arm. In the human the brachial and **basilic** veins unite to form the axillary. The tributaries to the axillary vein correspond, in general, to the branches of the axillary artery.

Major veins of the pectoral extremity

Deep veins These are the brachial vein and its tributaries, which parallel, in general, the brachial artery and its branches.

Superficial veins

Cephalic vein (Vena cephalica) You previously observed this vein on the dorsal side of the pectoral extremity when you removed the skin from the cat and when you studied the pectoral appendage. It joins the transverse scapular in the cat, but the axillary vein in the human.

Median cubital (Vena mediana cubiti) You observed this vein, too, when you studied the pectoral appendage. It is a communicating vein between the cephalic and brachial veins in the cat, and between the cephalic and basilic veins in the human. (The basilic vein of the human is located on the medial side of the extremity. It joins the brachial, as noted above, in formation of the axillary vein.)

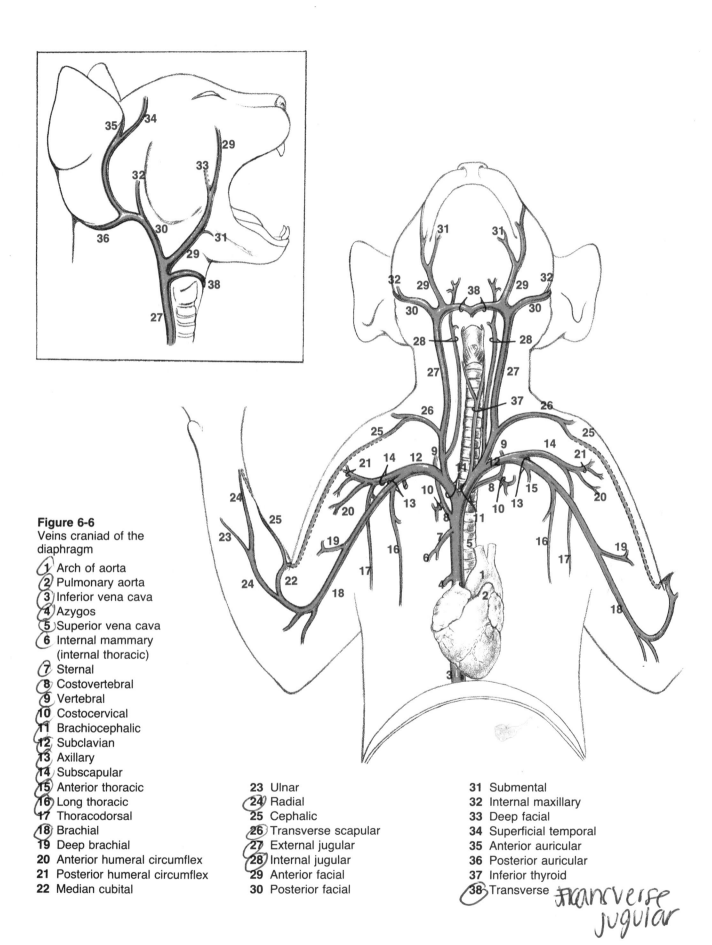

Figure 6-6
Veins craniad of the
diaphragm

1 Arch of aorta
2 Pulmonary aorta
3 Inferior vena cava
4 Azygos
5 Superior vena cava
6 Internal mammary
  (internal thoracic)
7 Sternal
8 Costovertebral
9 Vertebral
10 Costocervical
11 Brachiocephalic
12 Subclavian
13 Axillary
14 Subscapular
15 Anterior thoracic
16 Long thoracic
17 Thoracodorsal
18 Brachial
19 Deep brachial
20 Anterior humeral circumflex
21 Posterior humeral circumflex
22 Median cubital

23 Ulnar
24 Radial
25 Cephalic
26 Transverse scapular
27 External jugular
28 Internal jugular
29 Anterior facial
30 Posterior facial

31 Submental
32 Internal maxillary
33 Deep facial
34 Superficial temporal
35 Anterior auricular
36 Posterior auricular
37 Inferior thyroid
38 Transverse *Transverse*
jugular

Blood Vessels of the Brain

The arteries and veins of the brain will be considered in the chapter on the nervous system.

LYMPHATIC VESSELS
(Fig. 6-7)

You should read the chapter in your textbook dealing with the lymphatic system, and study the illustrations, before proceeding with the laboratory work.

The lymphatic system is not very satisfactory for dissection, but some of the major structures can be identified.

Throughout your dissection, you have noted various "lumps" of tissue which were identified as **lymph nodes.** These are found in various places throughout the body, including the neck region, axillary and popliteal fossae, the inguinal region, and in the mesentery. Small lymphatic vessels lead to (afferent vessels) and from (efferent vessels) the lymph nodes. You have observed some of the larger vessels, which often accompany veins. These are yellow-brown in color, are very thin-walled, and have a beaded appearance due to internal valve structures.

Below the diaphragm, the lymph stream is directed to a central receptacle, the **cisterna chyli,** which lies dorsal to the aorta just below the diaphragm. The cisterna chyli is difficult to observe because it is not desirable to destroy the aorta or the nerves and nerve ganglia in the region. At a later time, when you are studying the abdominal aorta, if you lift the aorta slightly you may be able to see the cisterna. The **thoracic duct,** one of the three major lymphatic vessels, arises from the cisterna chyli and proceeds craniad through the aortic hiatus (between the crura) of the diaphragm. It passes along the dorsal aspect of the aorta, first to the left and then veering to the right. Craniad of the aorta, it lies along the left side of the esophagus. It passes dorsal to the left subclavian artery, and then dorsal to the left brachiocephalic vein to reach the external jugular vein, which it joins (in the cat). The thoracic duct often splits into parallel vessels that course a short distance and then reunite.

The **right** and **left lymphatic trunks** are the other major vessels of this system. Each trunk, on its respective side, lies in the carotid sheath along with the common carotid artery, internal jugular vein, vagus nerve, and the cervical portion of the sympathetic trunk. The left lymphatic trunk joins the thoracic duct just before the latter joins the venous system, so the thoracic duct returns fluid from all levels caudal to the diaphragm and from the upper left side of the body. The right lymphatic trunk, which returns the fluid from the upper right side, joins the right external jugular vein.

In the human the thoracic duct, on the left, and the right lymphatic trunk typically join the internal jugular vein at its junction with the subclavian.

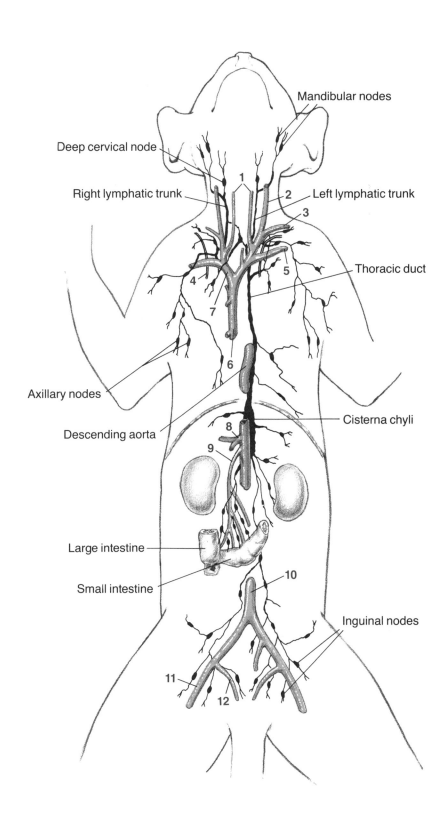

Mandibular nodes

Deep cervical node

Right lymphatic trunk

Left lymphatic trunk

Thoracic duct

Axillary nodes

Cisterna chyli

Descending aorta

Large intestine

Small intestine

Inguinal nodes

Figure 6-7
Lymphatic system

1 Internal jugular vein
2 External jugular vein
3 Transverse scapular vein
4 Thyrocervical artery
5 Subclavian vein
6 Superior vena cava
7 Brachiocephalic vein
8 Celiac artery
9 Superior mesenteric artery
10 Inferior vena cava
11 External iliac vein
12 Internal iliac vein

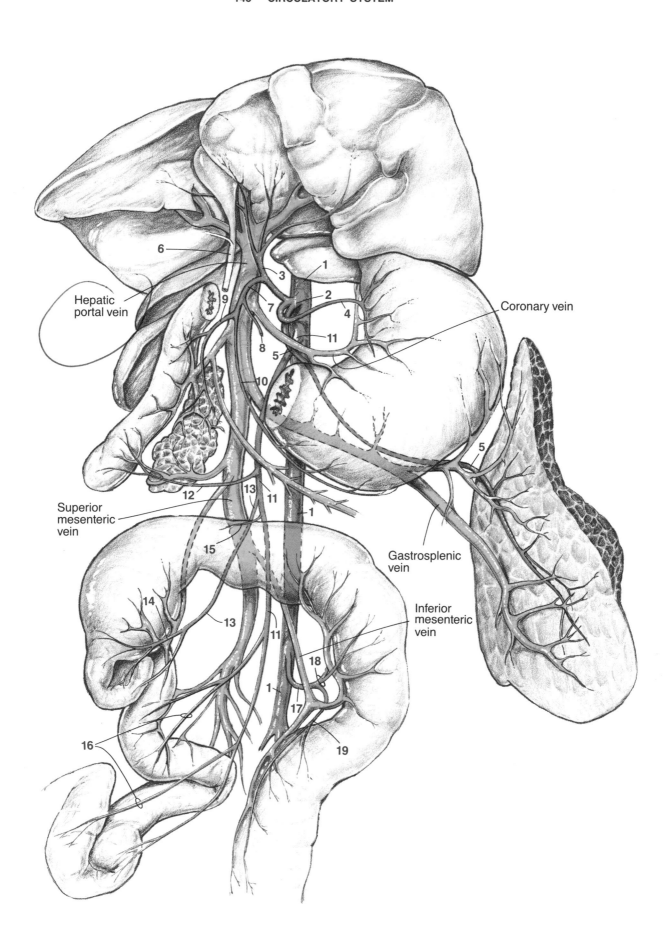

Hepatic
portal vein

Coronary vein

Superior
mesenteric
vein

Gastrosplenic
vein

Inferior
mesenteric
vein

VESSELS CAUDAD OF THE DIAPHRAGM

Arteries

(Figs. 6-8, 6-9, 8-18)

Refer to Figures 5-2 and 5-4 for some of the general relationships between the descending aorta and its branches and the inferior vena cava and its tributaries.

Locate the descending aorta as it passes through the aortic hiatus in the diaphragm, caudad of which it is known as the abdominal aorta. Since the vessels are outside of the peritoneum, this peritoneum must be at least partially dissected away with a probe—but *take great care* that nerves and nerve ganglia are not destroyed. Locate the arteries as indicated.

Visceral branches of the abdominal aorta

Celiac (coeliac) **trunk** An unpaired artery given off immediately caudad of the diaphragm. Gives the following branches that supply the stomach, liver, pancreas, duodenum, and spleen:

Left gastric To the lesser curvature of the stomach.

Hepatic To the liver and other viscera via its branches. The **gastro-duodenal** branch gives a **pyloric artery** to the lesser curvature side of the pylorus, a **right gastroepiploic artery** to the greater curvature side of the pylorus and to the omental bursa, and terminates as the **superior pancreaticoduodenal,** which courses to the duodenum and the head and neck of the pancreas. Other branches of the hepatic artery are a **cystic artery** to the gall bladder, and branches to lobes of the liver. (In the human the artery to the lesser curvature side of the pylorus branches directly from the hepatic, and is called the **right gastric.**)

Splenic To the greater curvature side of the stomach, the spleen, the body and tail of the pancreas, and the omental bursa.

Superior, or **anterior mesenteric** An unpaired artery supplying the head of the pancreas, small intestine, and large intestine as far caudad as the splenic flexure. It gives the following branches:

Inferior pancreaticoduodenal To the head of the pancreas and the duodenum.

Ileocolic To the caudal end of the ileum and to the cecum.

Right colic To the ascending colon. This artery may be a branch of the ileocolic.

Middle colic To the transverse colon.

Intestinals Course through the mesentery to the jejunum and the ileum.

Inferior, or **posterior mesenteric** An unpaired artery to the descending colon and rectum through the following branches:

Left colic To the descending colon.

Superior hemorrhoidal To the rectum. (In the human this artery also supplies the sigmoid colon.)

Note the anastomoses between the arteries of the digestive viscera, particularly between the intestinals, and between the arteries supplying the large intestine.

Figure 6-8
Arteries supplying the spleen and digestive viscera

1 Abdominal aorta
2 Celiac
3 Hepatic
4 Left gastric
5 Splenic
6 Common bile duct
7 Gastroduodenal
8 Pyloric
9 Superior pancreaticoduodenal
10 Right gastroepiploic
11 Superior mesenteric
12 Inferior pancreaticoduodenal
13 Ileocolic
14 Right colic
15 Middle colic
16 Intestinal
17 Inferior mesenteric
18 Left colic
19 Superior hemorrhoidal

Renal Paired vessel to the kidneys.

Internal spermatic (in the male), or **ovarian** (in the female) Paired vessel to the reproductive structures.

Adrenolumbar This paired vessel is a combination visceral and parietal branch to the adrenal glands and body wall. (In the human the artery to the adrenal gland is separate.)

Parietal branches of the abdominal aorta

Lumbar Several pairs supplying the body wall.

Iliolumbar Paired artery to the body wall. (In the human this is a branch of the internal iliac.)

Terminal branches of the aorta

External iliac Paired. It has the following branches:

Deep femoral Given off just before the external iliac passes into the thigh to continue as the femoral. A branch of the deep femoral artery, the **inferior epigastric,** supplies the ventral body wall; other branches supply the urinary bladder and external genitalia. (In the human the deep femoral is a branch of the femoral and the inferior epigastric is a branch of the external iliac.)

Internal iliac (hypogastric) Paired. Its branches go to pelvic viscera and to the gluteal region:

Umbilical To the urinary bladder.
Superior gluteal To dorsal hip muscles.
Middle hemorrhoidal To the caudal end of the rectum, urethra, and other tissues in the area. In the female a branch goes to the uterus, and in the male there are branches to the genital structures. (The uterine artery may be a branch of the internal iliac, as it typically is in the human.)
Inferior gluteal The terminal portion of the internal iliac. It supplies dorsal hip muscles.

Caudal, or **middle sacral** Unpaired. This small artery to the coccygeal region is the terminal portion of the aorta.

In the human the descending aorta gives off the **common iliac arteries** and then terminates as the small middle sacral artery. Each common iliac divides into an internal and an external iliac artery.

Major arteries of the pelvic extremity

Femoral The continuation of the external iliac artery. It gives off the following branches and continues as the popliteal artery:

Saphenous On the medial side of the extremity, along with the greater saphenous vein and saphenous nerve.
Branches to muscles

Figure 6-9
Arteries caudad of the diaphragm

1 Aorta (descending)
2 Celiac
3 Hepatic
4 Left gastric
5 Splenic
6 Cystic
7 Gastroduodenal
8 Pyloric
9 Right gastroepiploic
10 Superior pancreaticoduodenal
11 Superior mesenteric
12 Inferior pancreaticoduodenal
13 Middle colic
14 Right colic
15 Ileocolic
16 Intestinals
17 Adrenolumbar
18 Phrenic
19 Renal
20 Lumbars
21 Internal spermatic, or ovarian
22 Inferior mesenteric
23 Left colic
24 Superior hemorrhoidal
25 Iliolumbar
26 External iliac
27 Internal iliac (hypogastric)
28 Umbilical
29 Superior gluteal
30 Middle hemorrhoidal
31 Inferior gluteal
32 Caudal (middle sacral)
33 Deep femoral
34 Inferior epigastric
35 Branch to urinary bladder
36 Branch to external genitalia
37 Lateral femoral circumflex
38 Femoral
39 Branch to muscles
40 Saphenous
41 Popliteal
42 Posterior tibial
43 Anterior tibial

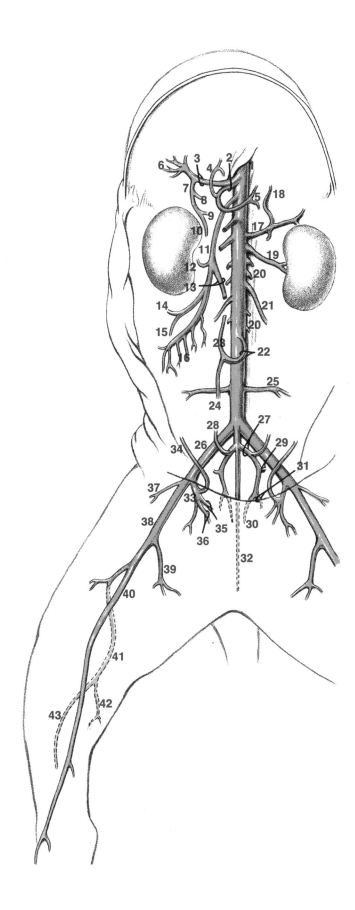

Popliteal The continuation of the femoral artery into the popliteal fossa. The popliteal artery gives off a rather large branch, the **sural** artery, which supplies parts of the biceps femoris and gastrocnemius muscles. Just proximal to the popliteus muscle, the popliteal artery divides into an anterior tibial and a posterior tibial artery.

Anterior tibial Passes ventrad of the popliteus muscle and through the interosseous membrane and courses through the leg and into the foot, giving off branches en route.

Posterior tibial Terminates in the dorsal crural muscles in the cat, but in the human it continues into the foot. In the human the posterior tibial gives off a **peroneal** branch, which courses down the lateral side of the leg and terminates in the heel region.

The tibial arteries give off branches in the foot that anastomose to form an arch (arches in the human), which in turn furnishes branches to the digits.

Veins
(Figs. 6-10, 6-11)

Major veins of the pelvic extremity

Deep veins These accompany the arteries in course and in name, and are the anterior and posterior tibial, popliteal, and femoral (and peroneal in the human).

Superficial veins You have already observed these veins, but you should review them at this point:

Greater saphenous Runs along the medial side of the extremity. Joins the femoral vein.

Lesser saphenous Runs along the dorsal side of the extremity. Joins a tributary of the internal iliac vein.

There are many anastomoses between the superficial veins, and between the superficial and the deep veins.

Veins of the abdominal region

Veins of this region, other than those from the digestive viscera and spleen, are the following:

External iliac A continuation of the femoral vein. It has tributaries paralleling the arteries, and it returns blood distributed by these arteries.

Internal iliac This vein, with its tributaries, returns blood distributed by the internal iliac artery and its branches. The internal iliac unites with the external iliac to form the common iliac vein.

Common iliac Formed by the union of the internal and external iliac veins. The **caudal vein** (middle sacral) joins either the right or the left common iliac vein.

Inferior vena cava (postcaval vein) Formed by the union of the common iliac veins. Its tributaries are the following:

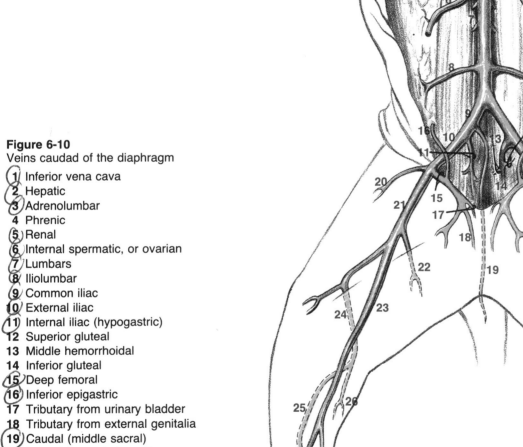

Figure 6-10

Veins caudad of the diaphragm

1 Inferior vena cava
2 Hepatic
3 Adrenolumbar
4 Phrenic
5 Renal
6 Internal spermatic, or ovarian
7 Lumbars
8 Iliolumbar
9 Common iliac
10 External iliac
11 Internal iliac (hypogastric)
12 Superior gluteal
13 Middle hemorrhoidal
14 Inferior gluteal
15 Deep femoral
16 Inferior epigastric
17 Tributary from urinary bladder
18 Tributary from external genitalia
19 Caudal (middle sacral)
20 Lateral femoral circumflex
21 Femoral
22 Tributary from muscles
23 Greater saphenous
24 Popliteal
25 Anterior tibial
26 Posterior tibial

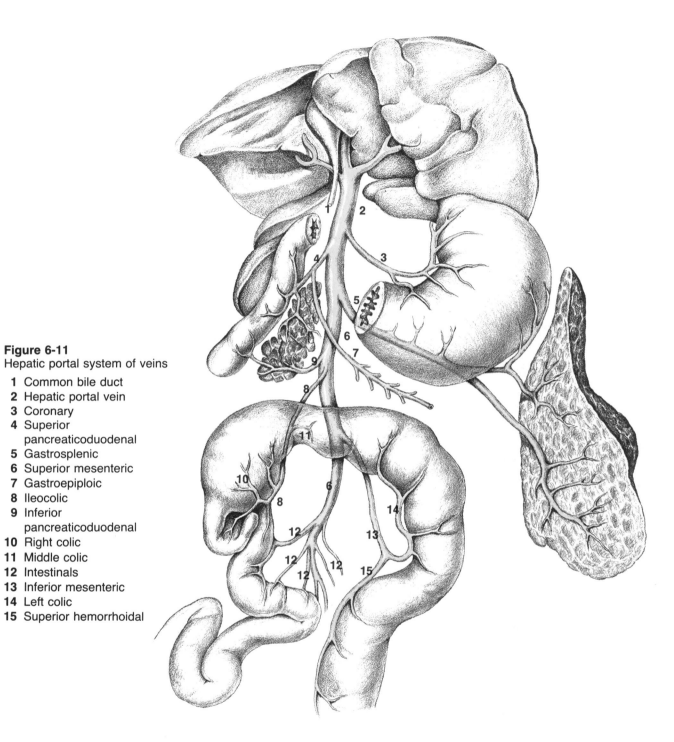

Figure 6-11
Hepatic portal system of veins

1 Common bile duct
2 Hepatic portal vein
3 Coronary
4 Superior
   pancreaticoduodenal
5 Gastrosplenic
6 Superior mesenteric
7 Gastroepiploic
8 Ileocolic
9 Inferior
   pancreaticoduodenal
10 Right colic
11 Middle colic
12 Intestinals
13 Inferior mesenteric
14 Left colic
15 Superior hemorrhoidal

Paired veins Correspond to the *paired* branches of the abdominal portion of the aorta and bear the same names: *iliolumbar, lumbar, renal, adrenolumbar,* and *internal spermatic* (or *ovarian*). The left internal spermatic (or ovarian) usually joins the left renal vein rather than the inferior vena cava.

Hepatic veins These begin in capillaries in the liver (liver, or hepatic sinusoids) and join the inferior vena cava as it passes through the liver. They are variable in number, but there are three in most humans.

The inferior vena cava has no tributaries craniad of the diaphragm, which it passes through on its course to the right atrium.

Hepatic portal system
(Figs. 4-5, 6-11)

The veins of the hepatic portal system will be found coursing through the mesentery and the greater and lesser omenta.

The unpaired **hepatic portal vein** receives blood from all the digestive viscera except the liver, and from the spleen, via veins corresponding to the arteries that distribute blood to these viscera. Find the veins, at least the major ones, using Figure 6-11 to determine the names and approximate locations.

The hepatic portal vein can be located in the lesser omentum, along with the common bile duct and the hepatic and gastroduodenal arteries. In the cat it is formed by the union of the **gastrosplenic vein** and the **superior mesenteric vein,** the later having tributaries that correspond to branches of the superior mesenteric artery. In the human the hepatic portal vein is formed by the union of the **splenic vein** and the superior mesenteric vein.

In the cat the **inferior mesenteric vein,** which has tributaries corresponding to branches of the inferior mesenteric artery, joins the superior mesenteric. In the human it joins the splenic vein.

The hepatic portal vein divides at the liver to give branches to each lobe that terminate in wide capillaries, called **liver,** or **hepatic sinusoids,** within the liver substance. The sinusoids are drained by the hepatic veins, which join the inferior vena cava. Blood carried to the liver by the hepatic artery passes through these same sinusoids.

7 ENDOCRINE GLANDS

ENDOCRINE GLANDS

The endocrine glands are those glands that secrete hormones. The glands do not have ducts and the hormones pass from the secreting cells into adjacent capillaries. You have already observed several of the glands described below; you will observe others when you study the brain.

Thyroid Gland
(Fig. 4-3)

The thyroid has two lobes, one on each side of the trachea just caudad of the larynx. The lobes are connected across the ventral surface of the trachea by a narrow band of glandular tissue called the **isthmus.** The isthmus is very small in the cat.

Parathyroid Glands

There are two pairs of parathyroid glands, one cranial and one caudal, lying against the dorsal surface of the lobes of the thyroid gland. These are too small for observation.

Thymus Gland
(Fig. 4-3)

This gland, which you have already removed, is present in varying degrees in the adult. It lies ventral to the trachea and may extend as far caudad as the heart.

Pancreas
(Figs. 4-4, 4-5)

You have already studied the pancreas (Chapter 4). The **islets of Langerhans,** which compose the endocrine portion, can be seen only in histological preparations.

Adrenal Glands
(Figs. 5-2, 5-4)

The adrenal glands (or suprarenals) are small oval bodies, one craniad and somewhat mediad of each kidney. (In the human the adrenals are situated, like caps, immediately over the cranial extremity of each kidney.) You observed these when you studied the urogenital system (Chapter 5).

Reproductive Glands
(Figs. 5-2, 5-4, 5-5,)

Certain cells of ovaries and testes are endocrine in function.

Pituitary Gland
(Figs. 4-1, 8-3, 8-5, 8-6)

To locate the pituitary gland (or hypophysis), make a sagittal section of the head, using a bone saw. (Upon occasion it may be desirable to remove the skull cap and lift the whole brain from the cranial cavity. If it is, you must do so before making the sagittal section. Directions for this procedure are given at the end of this chapter. After you have removed the brain, section the remainder of the head in the sagittal plane.) Saw through the cranium, and hard and soft palate, and the first two or three cervical vertebrae. *Try to leave the tongue intact*; this will be more likely if you stuff some cotton or paper into the oral cavity before making the section.

Pull the sections apart so that the epiglottis and the opening to the larynx are exposed. The pituitary can be observed projecting from the hypothalamus on the ventral side of the brain, within the sella turcica of the sphenoid bone. If the section is exactly through the midline—a difficult achievement—you will also have a section through the midline of the pituitary gland.

Pineal Gland
(Fig. 8-3)

This gland is quite small and was probably destroyed if the saw passed through it. If intact, it can be located at the midline, projecting from the caudal border of the roof of the diencephalon.

STUDIES WITH A SAGITTAL SECTION OF THE HEAD

Now that you have made a sagittal section of the head, you should observe the most cranial parts of the respiratory and digestive systems. For a list and description of these, refer to Chapter 4, pages 112–114.

DISSECTING FOR THE WHOLE BRAIN

Cut the muscle attachments from the first two or three neural arches and reflect the muscle. Cut the muscle attachments from the occipital and parietal bones, and reflect the muscle. Using bone shears, remove the neural arches to expose the spinal cord; then remove the dorsal, or superior portion of the occipital bone and a part of the lambdoidal ridge (between the parietal and occipital bones) to determine the position of the cerebellum. From this point, cut around the skull cap and lift it off. Note the dura mater of the brain. Carefully remove pieces of bone until the brain is sufficiently exposed to be lifted from the cranial cavity. A bony ledge that separates the cerebellum from the cerebral hemispheres is fused at its outer edge to the parietal bones, and this connection must be cut. To remove the brain from the cranial cavity, sever the spinal cord caudal to the medulla oblongata, sever the nerve and blood vessel connections, and lift the cord from the floor of the vertebral canal. Continue craniad, carefully working the brain loose from the floor of the cranial cavity, severing nerve and blood vessel connections as you proceed. Try to leave a small portion of the nerves and vessels attached to the brain for identification purposes. It will be difficult, if not impossible, to remove the hypophysis and olfactory bulbs intact.

Insofar as practical, the dura mater (p. 162) should be left intact, particularly in the floor of the cranial cavity, so that the dural sinuses (pp. 165–166) can be identified (possible only on well-injected specimens).

8 NERVOUS SYSTEM

The nervous system is made up of three divisions: the central nervous system (CNS), which consists of the brain and spinal cord; the peripheral nervous system (PNS), which consists of the cranial and spinal nerves; and the autonomic nervous system (ANS), which is a part of the other two divisions. This latter system furnishes the nerve supply for involuntary structures (visceral muscle, cardiac muscle, and glands).

We will begin our study with the brain and the cranial nerves.

THE BRAIN

In many classes the sheep brain will be used, rather than the cat brain, because it is larger and because of the time required to remove the cat brain from the cranial cavity. However, instructions for removal of the brain are given here, and the illustrations are based on the cat brain. These illustrations can also be used to study the sheep brain.

Remove the half section of brain* from one side of the head. Sever the spinal cord caudad of the medulla, and clip all nerves and blood vessels so that you leave a portion of them attached to the brain, if possible. Identify the structures indicated in the following descriptions.

---

*If you removed the whole brain (directions in Chapter 7) when you studied the endocrine glands, now study the brain as a whole, and then make a sagittal section of it when you are ready to study the ventricles.

Meninges

The meninges are the coverings of the brain and also of the spinal cord. They consist of three layers:

Dura mater The outer layer. This layer consists of two membranes: an outer periosteal, which is close to the inner surface of the skull and usually adherent to it, and an inner meningeal. The two membranes are closely applied to each other for the most part, but in some places there are certain venous channels, the dural sinuses, between the layers.

Arachnoid The middle layer, which is trabecular (cobweb-like) in structure. You may not be able to distinguish this layer.

Pia mater The inner layer, which closely invests the brain, dipping into the furrows.

In the living animal the spaces between the meningeal layers are filled with cerebrospinal fluid.

Consult your textbook for more detailed descriptions and for illustrations of the meninges.

Divisions of the Brain
(Figs. 8-1, 8-2, 8-3, 8-4, 8-6)

The brain has three primary divisions: **forebrain, midbrain,** and **hindbrain.**

Forebrain (prosencephalon) The forebrain consists of the following structures:

Telencephalon Includes the **cerebrum** and **olfactory bulbs.** The cat has larger olfactory bulbs relative to the size of the brain than either the sheep or the human, with the latter having the smallest ones. Note the two large **cerebral hemispheres,** which are characterized by convolutions (gyri) and furrows (sulci). The sulci and gyri of the cat brain have an alignment very much like those of the sheep. The gyri and sulci have names but it is beyond the scope of this manual to include these, particularly since the alignment is somewhat different in the human, and many of the names are not the same. Different lobes are recognized in a cerebral hemisphere, with the main ones being the frontal, the parietal, the temporal, and the occipital. Centers for reasoning, speech, smell, taste, vision, hearing, general sensory perception, and initiation of motor activity are localized in the various lobes. You should consult your textbook for the specific locations of these areas in the human brain.

In the cerebral hemispheres, the gray matter (composed mostly of nerve cell bodies and unmyelinated fibers) is on the outside and is called the **cerebral cortex.** The white matter (composed mostly of myelinated fibers) is internal. If you look down into the longitudinal fissure between the cerebral hemispheres, you can see the **corpus callosum,** a bridge of myelinated nerve fibers. These fibers carry impulses from one hemisphere to the other.

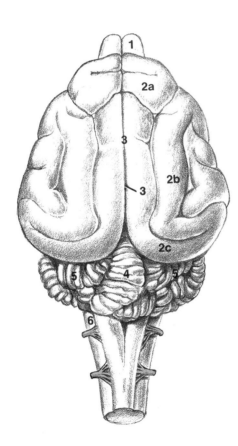

Figure 8-1
Brain, dorsal view

1 Olfactory bulb
2 Cerebral hemisphere
   a Frontal lobe
   b Parietal lobe
   c Occipital lobe
3 Longitudinal fissure
4 Vermis of cerebellum
5 Right and left hemispheres of cerebellum
6 Medulla oblongata

Figure 8-2
Brain, lateral view

 1 Olfactory lobe
 2 Olfactory tract
 3 Frontal lobe
 4 Parietal lobe
 5 Occipital lobe
 6 Temporal lobe
 7 Piriform lobe
 8 Cerebellum
 9 Pons
10 Trapezium
11 Medulla oblongata
12 Spinal cord
13 Optic nerve
14 Trigeminal nerve
15 Facial nerve
16 Abducens nerve
17 Acoustic nerve
18 Glossopharyngeal nerve
19 Vagus nerve
20 Spinal root of spinal
   accessory nerve
21 Hypoglossal nerve

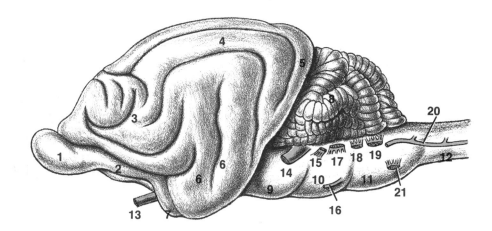

Diencephalon Composed mainly of the **thalamus** and **hypothalamus.** The hypophysis projects from the hypothalamus at the end of the **infundibular stalk.** Note the **optic chiasma** on the ventral surface, craniad of the hypophysis. The pineal gland, a part of the **epithalamus,** projects from the caudal border of the roof. The pineal gland of the sheep is larger than that of the cat or the human.

Midbrain (mesencephalon) The roof of the midbrain is composed of four bodies called the **corpora quadrigemina** (two **superior colliculi** and two **inferior colliculi**). The floor is composed of **cerebral peduncles,** one on each side, which are made up of ascending and descending nerve fibers, which are traveling between the cerebral hemispheres and lower levels.

Hindbrain (rhombencephalon) The hindbrain consists of the following:

Metencephalon Includes the **cerebellum** dorsally and the **pons** ventrally. Note the many folds of the cerebellum. The folds are much narrower than those of the cerebral hemispheres. The central portion of this foliate structure is called the **vermis,** with **right** and **left hemispheres** to either side. Here, as in the cerebral hemispheres, the gray matter is located on the external surface of the folds and is called the **cerebellar cortex.**

Note the ventral enlargement on the pons. This is caused mainly by nerve fibers passing transversely (from one side of the cerebellum to the other). The most caudal of the transverse fibers are in a part of the brain called the **trapezium,** or **trapezoid body.** In both the cat and the sheep brain this is usually considered to be a part of the medulla oblongata, but in the human there are more transverse fibers and a larger trapezium, which is included in the pons. The dorsal part of the pons is hidden by the overlying cerebellum.

Myelencephalon This is the **medulla oblongata.** Note its continuity with the spinal cord. The medulla oblongata, pons, and midbrain make up the **brain stem,** which connects the spinal cord with the forebrain.

Figure 8-3
Brain, sagittal section

 1 Olfactory bulb
 2 Cerebral hemisphere
 3 Corpus callosum
 4 Optic chiasma
 5 Infundibular stalk
 6 Hypophysis
 7 Massa intermedia of thalamus
 8 Superior colliculus of midbrain
 9 Pineal gland
10 Cerebral peduncle of midbrain
11 Pons
12 Medulla oblongata
13 Cerebellum
14 Spinal cord
15 Ventricle III
16 Cerebral aqueduct
17 Ventricle IV
18 Central canal of spinal cord

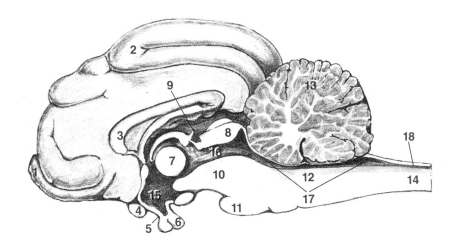

Ventricles of the Brain
(Fig. 8-3)

Within the brain there are a number of spaces that communicate with one another so that there is actually a continuous space, which communicates caudally with the canal of the spinal cord. The space is filled with cerebrospinal fluid in the living animal. The various spaces, from cranial to caudal, are the following:

Lateral ventricles The first two ventricles are the right and left ventricles of the telencephalon. Each ventricle communicates with ventricle **III** through an interventricular foramen, the **foramen of Monro.** (The lateral ventricles are not designated by number.)

Ventricle III In the diencephalon, mainly.

Aqueduct (cerebral aqueduct) A narrow canal through the midbrain. It connects the third and fourth ventricles. This canal is also called the **aqueduct of Sylvius.**

Ventricle IV In the hindbrain. It is continuous caudally with the **central canal** of the spinal cord.

Note the roof of the third ventricle in the diencephalic region and the roof of the fourth ventricle in the medulla oblongata. In these places there is only a thin membrane that contains tufts of tiny blood vessels, which produce the cerebrospinal fluid. This thin membrane is called the **tela choroidea,** and the blood vessels are called the **choroid plexus.** There are openings in the membranes that allow the fluid to get into the subarachnoid space.

Blood Vessels of the Brain
(Figs. 8-4, 8-5)

The rich supply of arteries to the brain can be observed on injected cats and in Fig. 8-4. In general, the arteries are named according to their location and the area they supply. They are continuations, or branches, of the vertebral, internal carotid, and internal maxillary arteries.

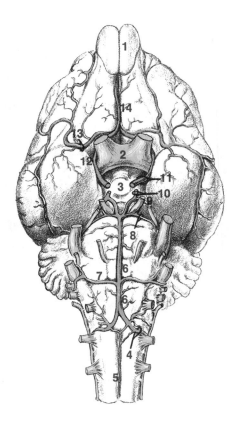

Figure 8-4
Brain, showing arteries,
ventral view

1 Olfactory bulb
2 Optic chiasma
3 Hypophysis
4 Vertebral artery
5 Anterior spinal artery
6 Basilar artery
7 Posterior inferior cerebellar
   artery

8 Anterior cerebellar artery
9 Posterior cerebral artery
10 Internal carotid artery (cut end)
11 Branch from internal maxillary
   artery (cut end)
12 Circle of Willis
13 Middle cerebral artery
14 Anterior cerebral artery

Note the formation of the **basilar** artery by union of the vertebral arteries, and the division of the basilar to form an **arterial circle** (of Willis) around the hypophysis and the optic chiasma. The internal carotid arteries and branches from the external carotid, through its internal maxillary extension, join the arterial circle.

The veins of the brain drain into the dural sinuses, which in turn empty into veins that join the internal jugular veins. The vertebral veins also receive blood returning from the brain, as well as from the spinal cord.

If you have left the dura mater in the cranial cavity when the brain was removed, and if the specimen is well-injected, the following dural sinuses can be observed (use Fig. 8-5 to aid in identification):

Superior sagittal Unpaired. In the dorsal portion of the falx cerebri (the dura mater in the longitudinal fissure).

Transverse In the dorsal border of the tentorium cerebelli (between the cerebellum and the cerebral hemispheres). In the human the tentorium is formed of dura mater; in the cat it is a shelf of bone covered by dura mater.

Inferior petrosal Along the medial border of the petrous portion of the temporal bone.

Cavernous and **intercavernous** Around the hypophysis and optic chiasma.

Vertebral In the floor of the neural canal, extending its entire length. The vertebral sinuses have connections with intercostal and lumbar veins, as well as with vertebral veins.

Communication between inferior petrosal and vertebral sinuses

Communication between the cavernous sinus and facial veins

There are additional small sinuses that do not usually receive an injection. Minor differences exist between the cat and the human, and you can consult a textbook for a detailed description and illustrations of the human dural sinuses.

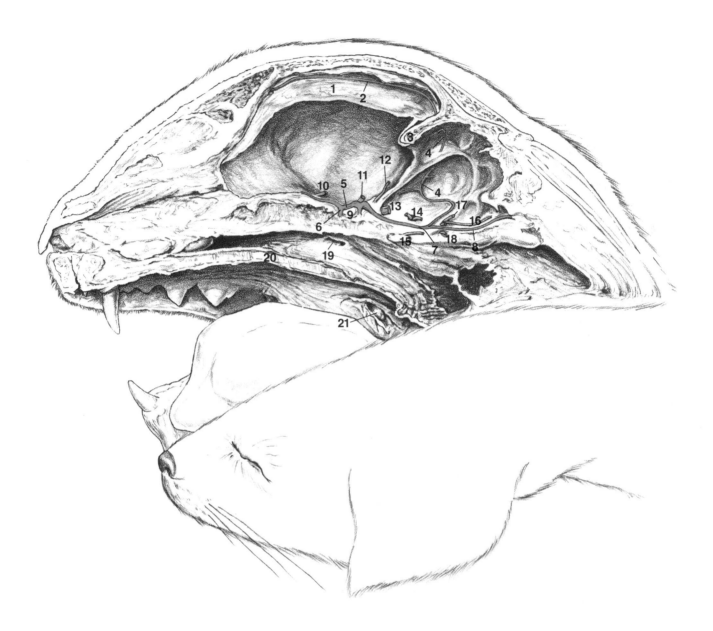

Figure 8-5
Dural sinuses of the cat

1 Falx cerebri
2 Superior sagittal sinus
3 Tentorium cerebelli
4 Transverse sinus
5 Cavernous sinus
6 Intercavernous sinus
7 Inferior petrosal sinus
8 Communication with
   vertebral sinus
9 Hypophysis lodged in
   sella turcica
10 Optic nerve
11 Oculomotor nerve

12 Trochlear nerve
13 Trigeminal nerve
14 Facial and acoustic nerves
   entering internal auditory
   meatus
15 Abducens nerve
16 Spinal root of accessory nerve
17 Glossopharyngeal and vagus
   nerves entering jugular foramen
18 Hypoglossal nerve
19 Opening of internal auditory tube
20 Soft palate
21 Palatine tonsil

CRANIAL NERVES

(Figs. 8-2, 8-5, 8-6)

Note any stumps of the cranial nerves that may have remained attached to the brain. These nerves pass through various foramina in the skull and distribute to head structures and some neck structures. (The vagus nerve passes farther caudad.)

The cranial nerves are frequently referred to by number rather than by name. Identify as many of the nerves as possible.

Olfactory (I) This is the nerve of smell and it consists of short processes that pass from the olfactory epithelium, in the roof of the nasal cavity, through the cribriform plate of the ethmoid bone to terminate in the **olfactory bulb.** Nerve fibers passing from the olfactory bulb to brain centers make up the **olfactory tract.**

Optic (II) This is the nerve of vision, with the cell bodies of the neurons being in the retina of the eye. The processes making up the nerve are axons of the third neurons in three-neuron sensory chains. The two optic nerves join in the optic chiasma, on the ventral surface of the hypothalamus, where the nerve fibers from the medial side of each retina cross to the opposite side. Caudad of the optic chiasma, these same tertiary-neuron axons compose the optic tract (on each side), which thus carries impulses from both eyes.

Oculomotor (III) Emerges from the ventral midbrain (from the cerebral peduncle). It supplies intrinsic eye muscles (the sphincter muscle of the pupil and the ciliary muscle) and four of the extrinsic eye muscles (the superior, medial, and inferior recti, and the inferior oblique). The oculomotor also supplies the levator palpebrae superioris muscle of the upper eyelid.

Trochlear (IV) Emerges from the roof of the midbrain at its caudal border, and supplies the superior oblique muscle (extrinsic eye muscle).

Trigeminal (V) Emerges from the lateral pons. There are three major divisions of this nerve, which may be identified if sufficient length of the nerve remains attached to the brain. These are (from dorsal to ventral) the ophthalmic, the maxillary, and the mandibular. The first two divisions are purely sensory, supplying general sensory fibers to the eyeball and the skin of the head and face, and will be mentioned again with the discussion of the orbit. The mandibular division provides general sensory supply for the skin of the lower face region, the teeth of the lower jaw, and the epithelium of the oral cavity and over the anterior two-thirds of the tongue. It also furnishes motor supply for the muscles of the mastication except for the posterior belly of the digastric (supplied by the facial nerve).

Abducens (VI) Emerges from the ventral pons at the caudal border, in the human, and supplies the lateral rectus muscle (an extrinsic eye muscle). In the cat, and also in the sheep, the abducens nerve emerges from the ventral medulla at the caudomedial border of the trapezium (you will recall that this region is a part of the pons in the human). In the cat there is an additional extrinsic eye muscle, the retractor oculi (retractor bulbi), supplied by the abducens.

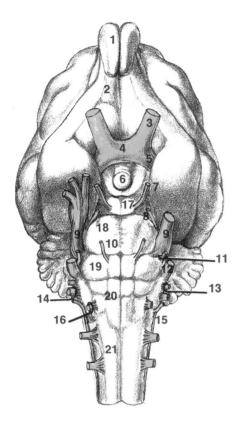

Figure 8-6
Brain, showing nerve attachments,
ventral view

 1 Olfactory bulb
 2 Olfactory tract
 3 Optic nerve
 4 Optic chiasma
 5 Optic tract
 6 Hypophysis
 7 Oculomotor nerve
 8 Trochlear nerve
 9 Trigeminal nerve
10 Abducens nerve
11 Facial nerve
12 Acoustic nerve
13 Glossopharyngeal nerve
14 Vagus nerve
15 Spinal root of accessory nerve
16 Hypoglossal nerve
17 Cerebral peduncle
18 Pons
19 Trapezium
20 Medulla oblongata
21 Spinal cord

Facial (VII) Emerges from the lateral pons, in the human. It supplies the cutaneous muscles of facial expression, and also the submandibular, sublingual, and minor salivary glands, the lacrimal gland, and the taste buds of the anterior two-thirds of the tongue. In the cat and the sheep the facial nerve emerges from the lateral trapezium.

Acoustic, or **auditory (VIII)** Joins the lateral pons at its junction with the medulla, in the human. The acoustic nerve has two roots, a cochlear and a vestibular. The vestibular portion supplies the semicircular canals and the vestibule of the inner ear (for equilibrium); the cochlear portion supplies the cochlea of the inner ear (for hearing). This nerve is also known as the **statoacoustic nerve,** and as the **vestibulocochlear nerve.** In the cat and the sheep the nerve emerges from the lateral trapezium.

Glossopharyngeal (IX) Emerges from the lateral medulla. It supplies muscles and epithelium of the pharynx, the parotid gland, and the epithelium of the posterior third of the tongue (for both taste and general sense).

Vagus (X) Emerges from the lateral medulla. It carries the motor supply for muscles of the larynx and for those of the pharynx not supplied by the glossopharyngeal nerve, and a sensory supply for the epithelium around the epiglottis and in the larynx, and perhaps some sensory supply for the skin around the auricula. It also carries both motor (parasympathetic) and sensory supply to thoracic viscera and to the abdominal viscera as far caudad as the splenic flexure of the colon.

Accessory, or **spinal accessory (XI)** Emerges from the lateral medulla. The spinal root supplies muscles of the neck (trapezius and sternocleidomastoid) and the cranial component distributes with the vagus nerve.

Hypoglossal (XII) Emerges from the ventral medulla. It supplies muscles of the tongue and some of the hyoid muscles.

THE ORBIT AND ITS CONTENTS
(Figs. 8-7, 8-8, 8-9, 8-10)

Remove the bony roof of the orbit with bone shears. Note the small superficial muscle, the **levator palpebrae superioris,** that operates the upper eyelid. Cut away the eyelids and free the eyeball from the border of the orbit except at the dorsomedial corner. Note the **lacrimal gland** on the dorsolateral surface of the eyeball. Cut away as much of the malar bone ventrally as is necessary to provide access to the orbit. Carefully pick away the fat and connective tissue (periorbital fascia), and separate the extrinsic muscles of the eye and blood vessels and nerves.

Various branches (ophthalmics) of the internal maxillary artery, and tributaries (ophthalmics) to the internal maxillary vein and to the deep facial and anterior facial veins will be found in the orbit.

Branches from the **ophthalmic** and **maxillary divisions** of the trigeminal nerve pass through the orbit. One branch from the ophthalmic division (branches of this division are dorsal to medial to the eyeball) supplies the general sensory fibers for the eyeball, but all of the others merely pass

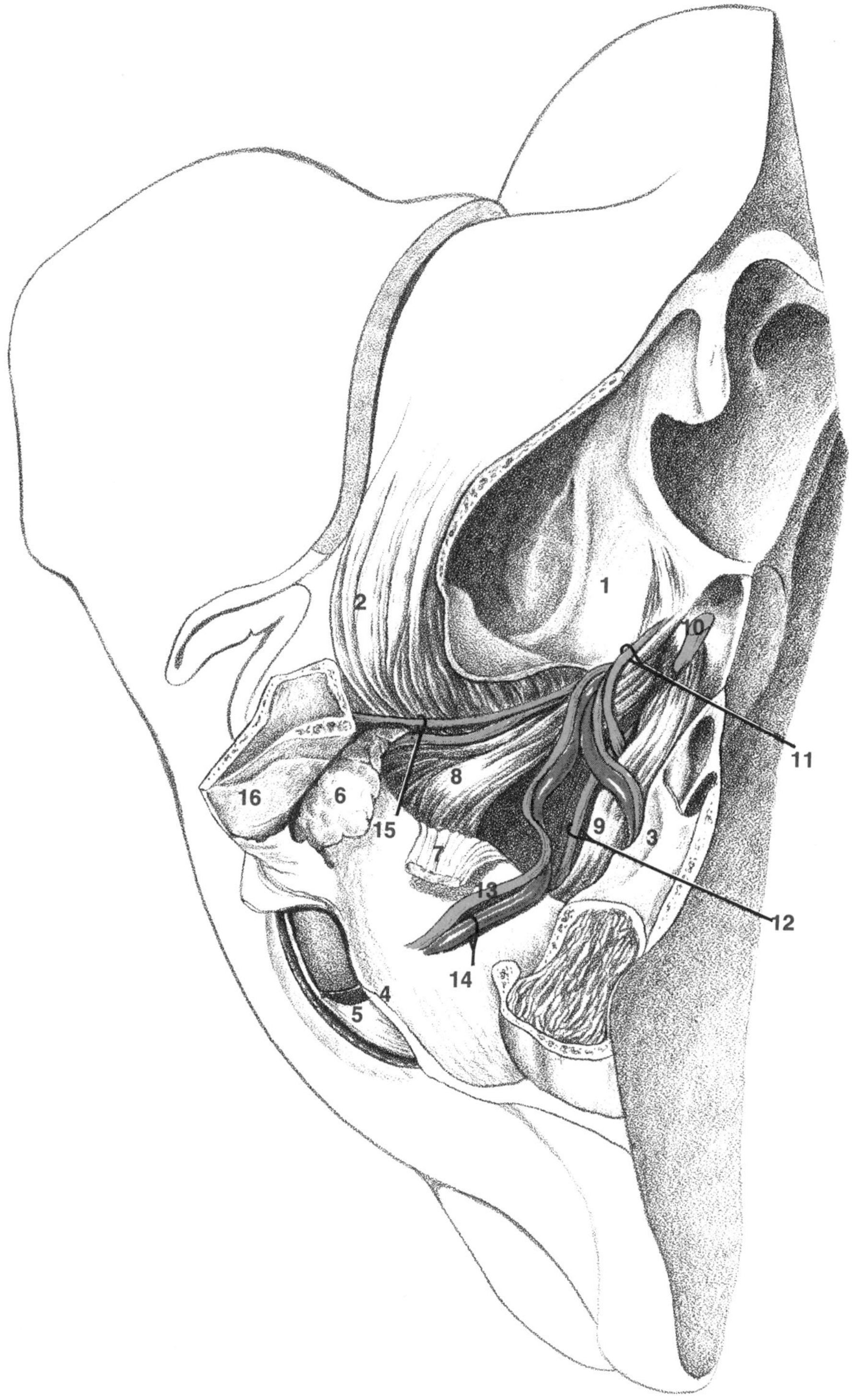

Figure 8-7
Contents of the orbit (right side),
dorsal view

 1 Cerebral fossa
 2 Temporalis muscle
 3 Medial wall of orbit
 4 Base of upper eyelid
 5 Nictitating membrane
 6 Lacrimal gland
 7 Levator palpebrae superioris
   muscle (cut near insertion)
 8 Superior rectus muscle
 9 Superior oblique muscle
10 Optic nerve (II)
11 Trochlear nerve (IV)
12 Infratrochlear nerve (branch of
   ophthalmic division of V)
13 Frontal nerve (branch of
   ophthalmic division of V)
14 Ophthalmic tributary to
   frontal vein
15 Lacrimal nerve (branch of
   maxillary division of V)
16 Inner surface of malar bone

through the orbit to supply the skin over the forehead, upper eyelid, and upper nasal area.

The **infraorbital** branch of the maxillary division of the trigeminal nerve passes across the floor of the orbit, lying on the external pterygoid muscle. (The cat does not have a complete bony orbit.) The infraorbital nerve gives off various branches that reach the epithelium of the oral cavity and the teeth of the upper jaw. (The teeth of the lower jaw are supplied by the **mandibular division** of the trigeminal nerve.) The infraorbital nerve passes through the infraorbital foramen to reach the skin of the face below the orbit and adjacent nasal area. Two other branches of the maxillary division, the **lacrimal** and **zygomatic,** course mediad of the temporalis muscle in the lateral wall of the orbit. The first passes by the lacrimal gland and leaves the orbit to supply the skin above the zygomatic arch; the second, which is more ventral in position, pierces the malar bone to reach the skin of the cheek area.

Locate the extrinsic muscles of the eye, which are given below. All except the inferior oblique arise at the apex of the orbit, which is marked by the optic foramen. (The cat has an additional extrinsic muscle, the **retractor oculi** or retractor bulbi. This muscle has four heads of insertion, all of which lie close to the eyeball and are partly covered by the other extrinsic muscles.)

Superior rectus Insertion on the dorsal surface of the eyeball. It lies deep to the levator palpebrae superioris.

Superior oblique Passes dorsally along the medial wall of the orbit and inserts on the dorsal surface of the eyeball (dorsolateral in the human). Note the connective tissue "pulley" in the dorsomedial corner, near the rim of the orbit, through which the tendon of insertion passes.

Medial rectus Insertion on the medial side of the eyeball, ventral to the superior oblique muscle.

Lateral rectus Insertion on the lateral side of the eyeball.

Inferior oblique Arises from the medial wall of the orbit and inserts on the ventrolateral surface of the eyeball.

Inferior rectus Insertion on the ventral surface of the eyeball.

The extrinsic eye muscles roll the eyeball in various directions. The superior rectus rolls it upward, the inferior rectus rolls it downward, the medial rectus rolls it mediad, and the lateral rectus rolls it laterad. The oblique muscles roll the eyeball either clockwise or counter-clockwise; the action will not be the same for a given oblique muscle on each eyeball and you can figure out the particular movement by determining the direction the eyeball would move when a given oblique muscle contracts.

The superior oblique muscle is supplied by cranial nerve IV (the trochlear), the lateral rectus by cranial nerve VI (the abducens, which also supplies the retractor oculi of the cat), and the other extrinsic muscles by cranial nerve III (the oculomotor). The cranial nerves enter each extrinsic muscle, except the inferior oblique, at a site near the origin of that muscle; the inferior oblique receives a branch from the oculomotor nerve that crosses the ventral surface of the eyeball to reach the muscle.

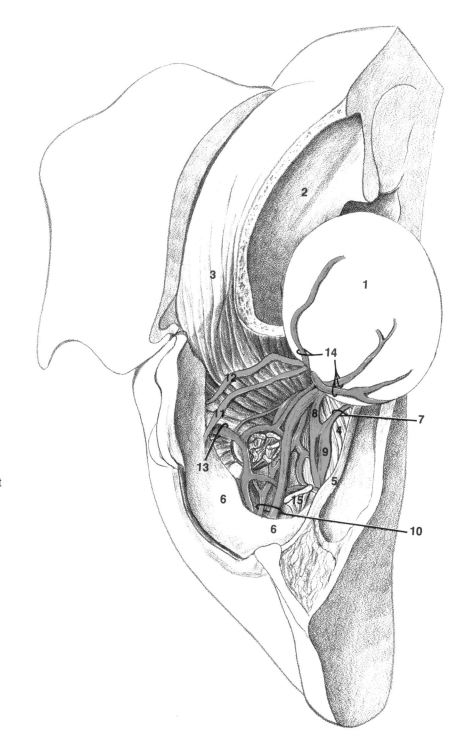

Figure 8-8
Structures in the floor of the orbit
(right side)

1 Eyeball reflected
2 Cerebral fossa
3 Temporalis muscle
4 External pterygoid muscle
5 Medial wall of orbit
6 Bony portion of floor of orbit
7 Vidian nerve
8 Sphenopalatine nerve
9 Sphenopalatine, or
   pterygopalatine ganglion
10 Infraorbital nerve (branch of
   maxillary division of V) and
   infraorbital artery
11 Zygomatic nerve (branch of
   maxillary division of V)
12 Lacrimal nerve (branch of
   maxillary division of V)
13 Anastomotic vein between
   superficial temporal and deep
   facial veins
14 Ophthalmic arteries and veins
15 Inferior oblique muscle at its
   origin

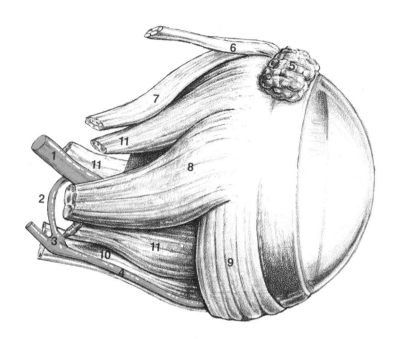

Figure 8-9
The eyeball (right), showing muscle attachments, lateral view

1 Optic nerve
2 Short ciliary nerve
3 Ciliary ganglion
4 Branch of oculomotor nerve (III) to inferior oblique muscle
5 Lacrimal gland
6 Levator palpebrae superioris muscle
7 Superior rectus muscle
8 Lateral rectus muscle
9 Inferior oblique muscle
10 Inferior rectus muscle
11 Retractor oculi muscle

The **ciliary ganglion** of the oculomotor nerve may be located lying on the inferior rectus muscle. This is a terminal autonomic ganglion* from which postganglionic fibers (parasympathetic) pass to two intrinsic eye muscles: the **ciliaris,** which operates the lens, and the **sphincter pupillae.** (The remaining intrinsic muscle, the **dilator pupillae** receives a sympathetic supply.) Another terminal autonomic ganglion, the **sphenopalatine** (pterygopalatine), lies medial to the infraorbital nerve. Parasympathetic preganglionic fibers from cranial nerve VII (the facial) course through the **Vidian** nerve, which is medial to the infraorbital, to synapse in the sphenopalatine ganglion, from which postganglionic fibers pass to the lacrimal gland and to minor salivary glands in the palate. Sympathetic postganglionic fibers also course in the Vidian nerve and pass to these same glands.

Note the optic nerve connection with the eyeball and its passage through the optic foramen.

Make a sagittal incision through the eyeball (see Fig. 8-10). (It is not necessary to remove the eyeball from the orbit unless your instructor directs you to.) Note the tunics of the eyeball: an outer fibrous **sclera;** a middle, highly vascular, thin **choroid** (which will be black or dark brown); and a thin inner **retina,** which contains nerve cells for vision. Note the translucent **cornea, crystalline lens,** and **iris.** The aperture in the iris is the **pupil.** The transparent epithelium that covers the exposed surface of the eyeball, and is continuous with epithelium of the eyelids, is the **conjunctiva.** The **ciliary body,** which rings the outer edge of the iris, is attached to the lens by a delicate suspensory ligament of radiating fibrils. This band of fibrils is called the **zonula ciliaris,** or **zonule of Zinn,** as well as the **suspensory ligament of the lens.**

*Some anatomists call this and other parasympathetic head ganglia *collateral* ganglia rather than terminal. The **otic** ganglion (which you will not see) that furnishes postganglionic neurons to the parotid gland is another such ganglion.

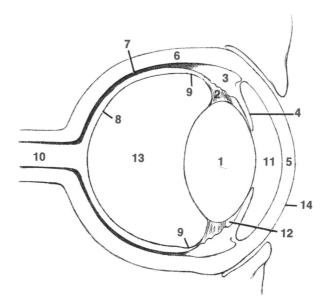

Figure 8-10
The eyeball, sagittal section

 1 Crystalline lens
 2 Suspensory ligament of lens
 3 Ciliary body
 4 Iris
 5 Cornea
 6 Sclera
 7 Choroid
 8 Retina
 9 Ora serrata of retina
10 Optic nerve
11 Anterior chamber
12 Posterior chamber
13 Vitreous body
14 Conjunctiva

Note the chambers of the eye. The **anterior chamber,** between the iris and the cornea, and the **posterior chamber,** bounded by the iris and by the lens and zonula ciliaris, contain a fluid called **aqueous humor** in the living animal. The cavity behind the lens and zonula ciliaris, is filled with the viscous **vitreous body** in the living animal.

THE VAGUS NERVE AND SOME OTHER CRANIAL NERVES

Locate the vagus nerve (Fig. 8-11) in the cervical region, where it will be found alongside the cervical portion of the sympathetic trunk, common carotid artery, and internal jugular vein (refer also to Fig. 6-4).

Note the large **nodose ganglion** (also called the **inferior ganglion,** since the vagus has another superior, or jugular ganglion) laterad, and slightly craniad, of the thyroid cartilage. It is laterad of, and closely bound to the superior cervical ganglion of the sympathetic trunk. It may be impossible to separate the two ganglia. Note the **superior laryngeal** branch to the larynx. The **inferior,** or **recurrent laryngeal** branch loops around the subclavian artery on the right, but around the arch of the aorta on the left, and ascends on the surface of the trachea to reach the larynx, where it supplies motor fibers to the muscles. As the vagus nerve courses caudad, it sends preganglionic (parasympathetic) and sensory fibers to all major thoracic viscera.

Caudad of the bronchus, the vagus divides into ventral and dorsal portions. The ventral divisions from each side unite to form the **ventral vagal trunk,** which courses caudad ventral to the esophagus and passes through the diaphragm to reach the lesser curvature of the stomach, which it supplies. The dorsal divisions from each side course caudad and unite to form the **dorsal vagal trunk,** which continues caudad dorsal to the esophagus and passes through the diaphragm to reach the greater curvature of the stomach, which it supplies. Preganglionic parasympathetic and sensory

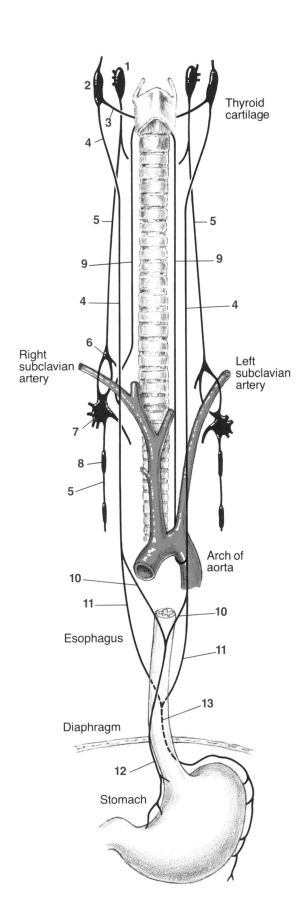

Figure 8-11
Vagus nerve and sympathetic trunk

  1 Superior cervical ganglion
  2 Nodose ganglion (inferior
     ganglion of vagus)
  3 Superior laryngeal nerve
  4 Vagus nerve
  5 Sympathetic trunk
  6 Middle cervical ganglion
  7 Stellate ganglion
  8 Chain ganglion
  9 Recurrent laryngeal nerve
     (inferior laryngeal)
10 Vagus nerve, ventral division
11 Vagus nerve, dorsal division
12 Ventral vagal trunk
13 Dorsal vagal trunk

Figure 8-12
Some superficial branches of the facial and trigeminal nerves

1 Masseter muscle
2 Lymph node
3 Submandibular gland
4 Sublingual gland
5 Parotid gland
6 Parotid duct
7 External jugular vein
8 Transverse vein
9 Posterior facial vein
10 Anterior facial vein
11 Dorsal ramus of facial nerve
12 Zygomatic branch of facial nerve
13 Fibers from dorsal ramus of facial nerve and from auriculotemporal nerve (a branch from mandibular division of trigeminal nerve)
14 Zygomatic branch from maxillary division of trigeminal nerve
15 Ventral ramus of facial nerve
16 Superior and inferior buccal nerves

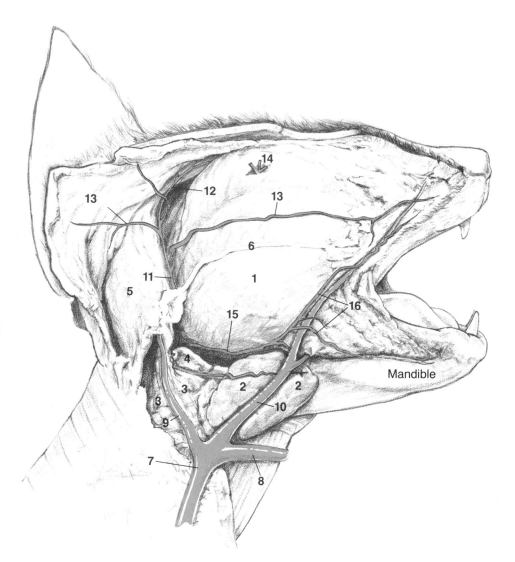

fibers, primarily from the dorsal trunk, are distributed to major abdominal viscera as far caudad as the splenic flexure of the colon. (Viscera caudad of the splenic flexure receive a parasympathetic supply from sacral spinal nerves.)

There are some other cranial nerves, or branches, which you can easily locate in the head and cervical regions. The hypoglossal nerve accompanies the lingual artery to the tongue. The accessory nerve pierces the cleido-mastoid muscle, which it supplies, and sends branches to the sternomastoid and to parts of the trapezius. The glossopharyngeal nerve lies immediately ventral to the skull and passes to the pharynx and to epithelium of the posterior third of the tongue.

The ventral and dorsal rami of the facial nerve are easily located. The ventral ramus is superficial to the lower border of the masseter muscle (craniad of the submandibular and sublingual glands), and the dorsal ramus is immediately in front of the external ear, deep to the parotid gland (see Fig. 8-12). The ventral ramus sends branches to the superficial muscles around the mouth; the dorsal ramus to those around the ear and eye, and the cheek. Other superficial branches of the facial nerve go to the posterior part of the digastric muscle and to superficial muscles on the dorsal surface of the skull. To follow the branches of the facial nerve to their points of origin, cut the parotid duct, and loosen the parotid gland and reflect it caudad. Find the point of union of the dorsal and ventral rami and then trace the nerve to its emergence from the skull through the stylomastoid foramen, just caudal to the external ear opening.

Some of the other superficial nerves that you encountered when removing the skin from the face and head are branches from the trigeminal nerve, which supplies the skin of the face and much of the scalp.

THE SPINAL CORD AND SPINAL NERVES
(Fig. 8-13)

You will not dissect the spinal cord, but you will study the nerves that are connected with it. The human has 31 pairs of spinal nerves: 8 cervical, 12 thoracic, 5 lumbar, 5 sacral, and 1 caudal or coccygeal. The cat has 38 or 39: 8 cervical, 13 thoracic, 7 lumbar, 3 sacral, and 7 or 8 caudal or coccygeal. A spinal nerve is formed by the union of a **ventral root** (motor) and a **dorsal root** (sensory); on the dorsal root is located a group of sensory nerve cell bodies which form an enlargement called the **dorsal root ganglion,** or **spinal ganglion.** The spinal nerve formed from the two roots soon divides into a **dorsal ramus** and a **ventral ramus.** Dorsal rami supply the deep muscles of the back and neck and the skin of the dorsal body wall. Ventral rami distribute to the superficial muscles of the back and neck, to muscles and skin of the ventral and lateral body wall, and to the appendages.

With some exceptions, the spinal nerves exit from the vertebral column through intervertebral foramina formed by apposition of two vertebrae. The first seven cervical nerves exit craniad of the cervical vertebra of the same number. The first cervical spinal nerve, then, exits between the atlas and the occipital bone. The eighth cervical nerve has its exit caudal to the seventh cervical vertebra. The rest of the spinal nerves exit caudal to the vertebra of the same name and number as the nerve.

The ventral ramus of the first thoracic spinal nerve sends a large branch ventrad and craniad of the first rib to join the brachial plexus, but the intercostal part of the ramus passes caudal to the rib, as do other intercostal nerves and the one subcostal nerve (caudal to the last rib).

In the sacral region, where the vertebrae are fused, the ventral rami of the sacral nerves pass through the anterior (ventral) sacral foramina and the dorsal rami through the posterior (dorsal) sacral foramina, except for the last sacral nerve, where both rami pass between the sacrum and the first coccygeal vertebra.

There is only one coccygeal, or caudal nerve in the human, and it exits caudal to the first coccygeal vertebra.

Associated with each spinal nerve is a group of motor nerve cell bodies that form a ganglion. These ganglia, called "chain ganglia," are interconnected by ascending and descending neuron processes, so that a continuous trunk, the **sympathetic trunk,** is formed. Some of these neuron processes

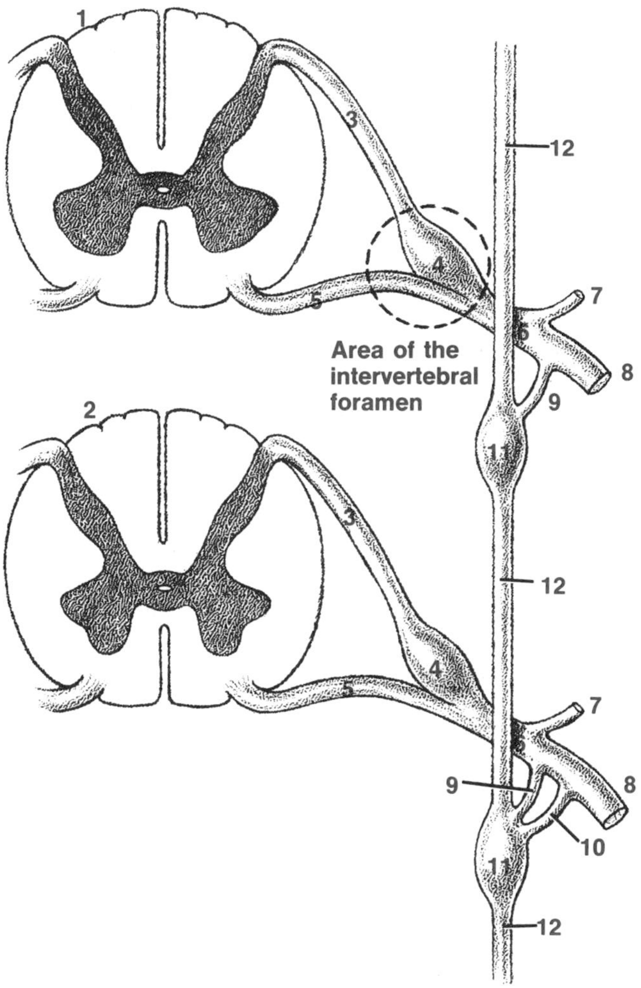

Figure 8-13
Relationship of spinal nerves to spinal cord and sympathetic trunk

 1 Represents levels of the spinal cord having no white ramus communicans
 2 Represents levels of the spinal cord having both gray and white rami communicantes
 3 Dorsal root of the spinal nerve
 4 Dorsal root ganglion
 5 Ventral root of the spinal nerve
 6 Trunk of the spinal nerve
 7 Dorsal ramus
 8 Ventral ramus
 9 Gray ramus communicans
10 White ramus communicans
11 Chain ganglion
12 Sympathetic trunk

Area of the intervertebral foramen

are those of postganglionic neurons with cell bodies in the chain ganglia; some are those of preganglionic neurons, with cell bodies in the spinal cord, that will synapse in chain ganglia at levels that are higher or lower than the point of emergence of the processes from the spinal cord; some are those of sensory neurons from major viscera. Only thoracic and upper lumbar nerves contribute preganglionic processes to the sympathetic trunk, but all spinal nerves have chain ganglia associated with them, and all contribute postganglionics.

THE SYMPATHETIC TRUNK AND ITS BRANCHES
(Figs. 8-11, 8-13, 8-14)

Locate the sympathetic trunk in the cervical region, where it is closely adjacent to the vagus nerve and is bound to it by connective tissue, so that it appears to be a part of the vagus.

Locate the three sympathetic ganglia that are fusions of some of the original separate ganglia. The **superior cervical ganglion** is medial to the nodose ganglion. This is a combination of the original chain ganglia of the

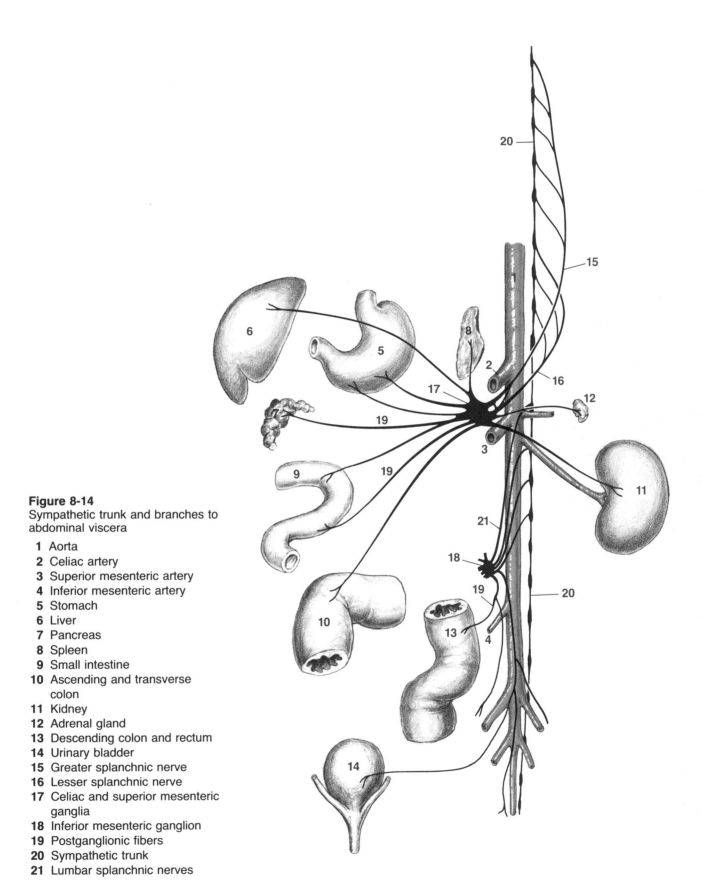

Figure 8-14
Sympathetic trunk and branches to
abdominal viscera

 1 Aorta
 2 Celiac artery
 3 Superior mesenteric artery
 4 Inferior mesenteric artery
 5 Stomach
 6 Liver
 7 Pancreas
 8 Spleen
 9 Small intestine
10 Ascending and transverse
    colon
11 Kidney
12 Adrenal gland
13 Descending colon and rectum
14 Urinary bladder
15 Greater splanchnic nerve
16 Lesser splanchnic nerve
17 Celiac and superior mesenteric
    ganglia
18 Inferior mesenteric ganglion
19 Postganglionic fibers
20 Sympathetic trunk
21 Lumbar splanchnic nerves

first four cervical nerves. Postganglionic processes leave the ganglion and follow the common carotid artery and its branches to the head region. The **middle cervical ganglion,** which is just craniad of the subclavian artery, is very small. This ganglion represents a fusion of the original chain ganglia of the fifth and sixth cervical nerves. Caudad of the middle cervical ganglion, the sympathetic trunk splits and forms a loop (ansa) around the subclavian artery. It then joins the **stellate ganglion,** which lies laterad of the vertebral bodies between the first and second ribs. The stellate ganglion represents fused chain ganglia of the seventh and eighth cervical nerves and the first four thoracic nerves. (In the human the ganglia of the seventh and eighth cervical nerves fuse to form an inferior cervical ganglion, which may or may not fuse with one or more thoracic chain ganglia to form a stellate ganglion.)

All three of these sympathetic ganglia give off nerves to the heart: The superior cervical ganglion gives off the **superior cardiac nerve**; the middle cervical ganglion, the **middle cardiac nerve**; the stellate ganglion, the **inferior cardiac nerve.** They also give off fibers to other thoracic viscera.

Caudad of the stellate ganglion the sympathetic trunk lies laterad of the bodies of the vertebrae, and small chain ganglia can be observed at intervals. Before passing through the diaphragm, the sympathetic trunk gives off the **greater** and **lesser splanchnic nerves,** which contain preganglionic processes that synapse in collateral, or intermediate, ganglia. These ganglia are the **celiac,** near the base of the celiac artery, and the **superior mesenteric,** near the base of the superior mesenteric artery. The ganglia are frequently fused, and those of the right and left sides are united by interconnecting fibers. Postganglionic processes leave the ganglia to follow the arteries and their branches to the major viscera.

Preganglionic fibers are given off from the lumbar portion of the sympathetic trunk and course caudad, as **lumbar splanchnic nerves,** to synapse in the **inferior mesenteric ganglion,** a collateral ganglion near the base of the inferior mesenteric artery. As with the other collateral ganglia, there are interconnecting fibers between the right and left sides. Postganglionics follow the artery and its branches to the major viscera. Postganglionics also follow the aorta and its terminal branches.

Postganglionic fibers are given off by all of the chain ganglia to supply minor body wall viscera at each level. These fibers traverse the gray ramus of the respective spinal nerve (the nerve furnishing the postganglionic neuron) and distribute through dorsal and ventral rami.

Note that the sympathetic trunk, as it courses caudad, swings mediad and occupies a position ventral to the vertebral bodies before its termination in the sacral region.

THE BRACHIAL PLEXUS
(Figs. 8-15, 8-16)

The brachial plexus is formed by the ventral rami of the last four cervical nerves and a large branch of the ventral ramus of the first thoracic nerve. The peripheral nerves arising from this plexus, which is located in the axilla, supply the pectoral appendage. Some of the nerves you observed previously.

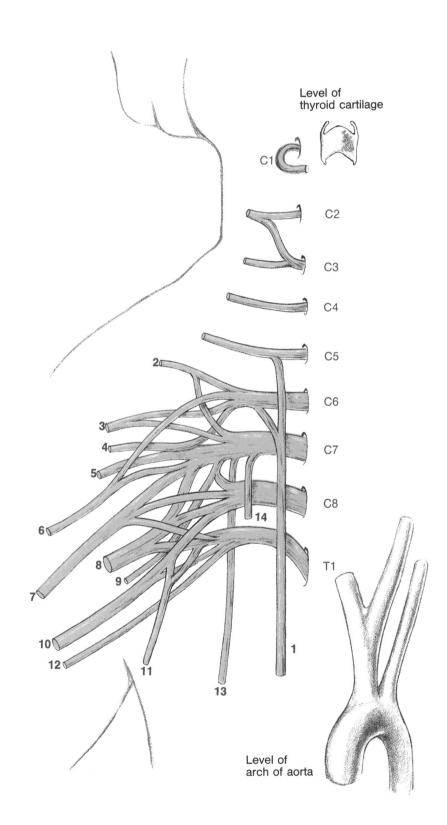

Figure 8-15
Nerves of the brachial plexus

  1 Phrenic nerve
  2 Suprascapular
  3 First subscapular
  4 Axillary
  5 Second subscapular
  6 Musculocutaneous
  7 Median
  8 Radial
  9 Third subscapular
10 Ulnar
11 Second anterior thoracic
12 Medial cutaneous
13 Posterior thoracic (long
     thoracic)
14 First anterior thoracic

Locate the following nerves, using Figures 8-15 and 8-16 as aids:

Phrenic Near its origin this nerve accompanies the thyrocervical artery. It passes ventral to the root of the lung and continues caudad to the diaphragm. The phrenic nerve is not actually a part of the brachial plexus. In the human it arises from a small cervical plexus; it is mainly from the fourth cervical nerve, but receives contributions from the third and fifth.

Suprascapular Passes over the cranial border of the scapula. Follows the transverse scapular artery to the supraspinatus and infraspinatus muscles.

First subscapular Follows the subscapular artery to the subscapularis muscle.

Second subscapular Goes to the subscapularis and teres major muscles.

Axillary Follows the posterior humeral circumflex artery to the teres minor and deltoid muscles.

Musculocutaneous Goes to the ventral muscles of the arm, and gives a cutaneous branch to the forearm.

Posterior thoracic, or **long thoracic** Goes to the serratus anterior and courses along the surface of that muscle.

First anterior thoracic Accompanies the anterior thoracic artery to the pectorales.

Third subscapular Follows the thoracodorsal artery to the latissimus dorsi.

Radial Crosses the ventral surface of the teres major muscle and passes dorsal to the humerus, along with the deep brachial artery, where it gives off branches to the dorsal arm muscles. If you transect the lateral head of the triceps brachii, you can observe the radial nerve beneath it (on the lateral surface of the brachialis muscle). Here it gives off a large cutaneous branch that passes into the forearm, along with the brachioradialis muscle, and into the hand. The radial nerve passes into the dorsal forearm muscles, all of which it supplies.

Median Supplies the ventral muscles of the forearm, except for the flexor carpi ulnaris and the ulnar head of the flexor digitorum profundus; supplies also some of the intrinsic hand muscles. In the cat this nerve accompanies the brachial and radial blood vessels.

Ulnar Passes dorsal to the medial epicondyle of the humerus and supplies the flexor carpi ulnaris and the ulnar head of the flexor digitorum profundus. It also supplies some of the intrinsic hand muscles. In the forearm, you will find this nerve near the ulnar artery.

Second anterior thoracic Accompanies the long thoracic artery to the latissimus dorsi and pectorales muscles.

Medial cutaneous Passes distad on the medial side of the arm to the skin on the medial side of the forearm.

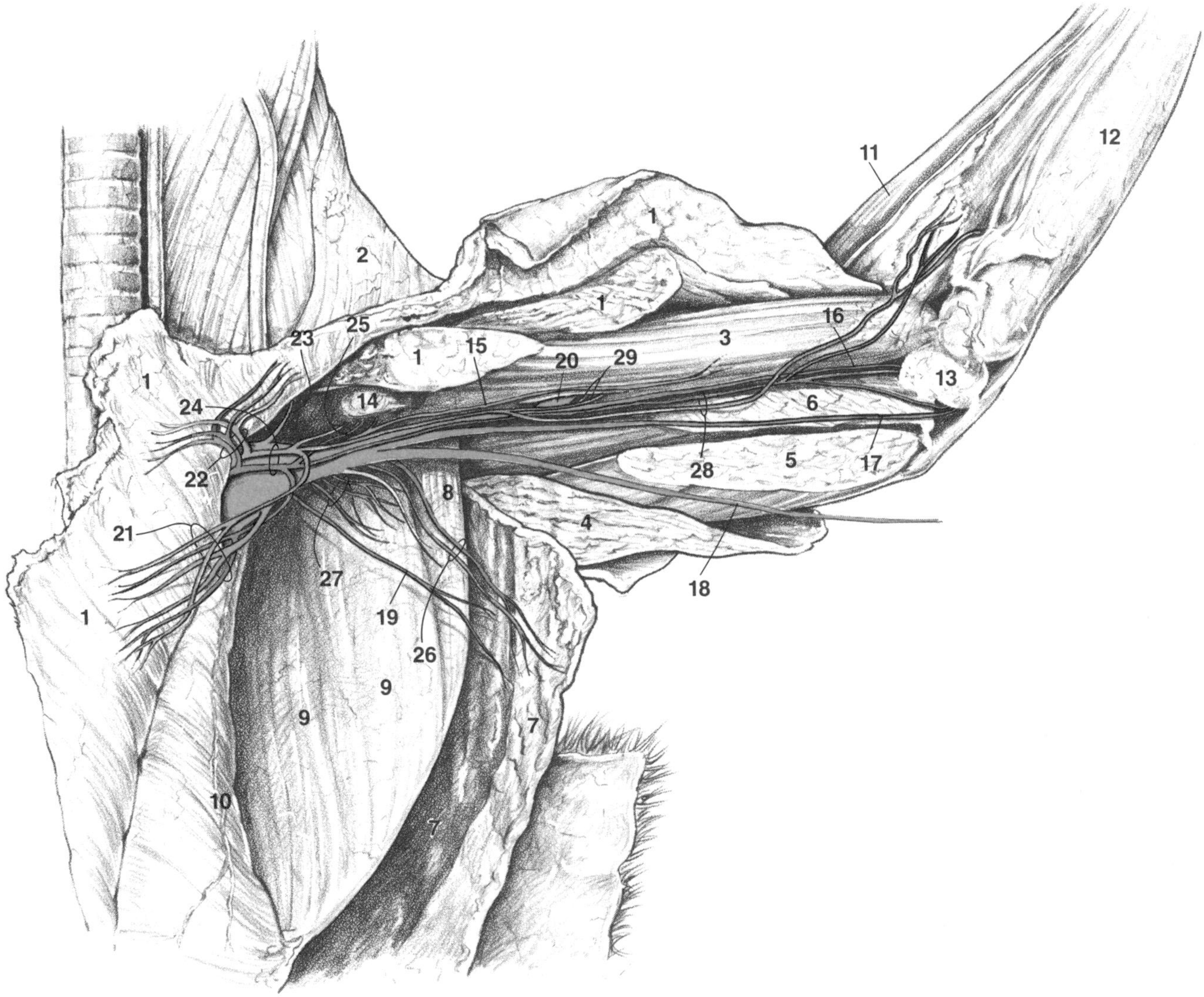

Figure 8-16
Some nerves of the brachial plexus, with associated vessels and muscles

1 Pectoralis muscles
2 Clavotrapezius
3 Biceps brachii (long head)
4 Epitrochlearis
5 Triceps brachii (long head)
6 Triceps brachii (a division of medial head)
7 Latissimus dorsi
8 Teres major
9 Subscapularis
10 Serratus anterior with posterior thoracic nerve embedded in fascia

11 Brachioradialis
12 Antebrachial fascia
13 Location of medial epicondyle of humerus
14 Coracobrachialis
15 Musculocutaneous nerve
16 Median nerve
17 Ulnar nerve
18 Medial cutaneous nerve
19 Third subscapular nerve
20 Radial nerve
21 Second anterior thoracic nerve and long thoracic vessels

22 First anterior thoracic nerve and anterior thoracic vessels
23 Suprascapular nerve
24 Axillary artery and vein
25 Subscapular artery and vein
26 Thoracodorsal artery and vein
27 First and second subscapular nerve and subscapular vessels
28 Brachial artery and vein
29 Deep brachial artery and vein

THE LUMBOSACRAL PLEXUS
(Figs. 8-17, 8-18)

The lumbosacral plexus is formed by ventral rami of lumbar and sacral spinal nerves. The nerves that contribute to the plexus vary slightly between the cat and the human. (The human has five pairs of lumbar and five pairs of sacral spinal nerves, whereas the cat has seven lumbar and three sacral.) The peripheral nerves arising from the plexus, which is located in the lower abdominal and pelvic regions, supply the skin and muscles of the pelvic region and the pelvic appendage. Some of the nerves you have observed previously.

Locate as many of the following nerves as possible, using Figures 8-17 and 8-18 as aids:

Genitofemoral Supplies the skin over the external genitalia, and the medial side of the thigh and adjacent abdominal body wall. This nerve has two branches, a medial and a lateral. The medial branch courses along the medial side of the iliopsoas and then accompanies the external iliac artery and its deep femoral branch. The lateral branch pierces the psoas minor and courses caudad ventral to the muscle, crosses the iliolumbar blood vessels, and proceeds caudad to its distribution.

Lateral femoral cutaneous Supplies the skin covering the lateral surface of the thigh and hip region. This nerve will be found close to the iliolumbar blood vessels.

Femoral Emerges from the psoas major and distributes to the ventral femoral muscles. It gives off a cutaneous branch, the **saphenous nerve,** which accompanies the greater saphenous vein and saphenous artery, to the leg and foot. You have already observed this nerve in the femoral triangle, laterad of the femoral artery. The femoral nerve also supplies the iliacus and pectineus muscles.

Obturator Passes through the obturator foramen and distributes to the medial femoral muscles, except the pectineus, and to the obturator externus. Just laterad of the pubic symphysis, you can find a branch of this nerve to the gracilis muscle; it emerges from under the adductor longus muscle and passes between gracilis and the adductor femoris muscle.

Just distal to the origin of the obturator nerve, and dorsal to the iliac blood vessels, you will observe a large nerve cord. This is the **lumbosacral cord,** or **trunk**; it is composed of portions of ventral rami of the sixth and seventh lumbar nerves, which join the ventral rami of sacral nerves. (Since the human has only five pairs of lumbar nerves, the lumbosacral cord is composed of a branch of the ventral ramus of the fourth lumbar and the ventral ramus of the fifth lumbar nerve.)

Superior gluteal Can be found accompanying the superior gluteal blood vessels. It supplies the gluteus medius, gluteus minimus, and tensor fasciae latae muscles. Branches of the nerve can be observed along with the blood vessels, deep between opposing margins of the gluteus medius and tensor fasciae latae.

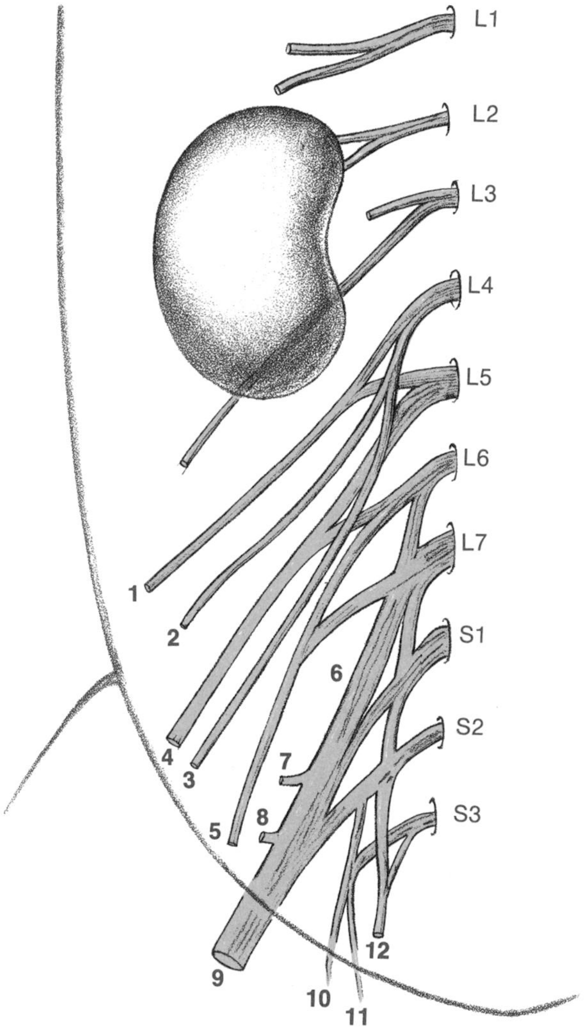

Figure 8-17
Nerves of the lumbosacral plexus

  1 Lateral femoral cutaneous
  2 Genitofemoral, lateral branch
  3 Genitofemoral, medial branch
  4 Femoral
  5 Obturator (The obturator nerve
    may arise from L5 and L6,
    rather than L6 and L7.)
  6 Lumbosacral cord
  7 Superior gluteal
  8 Inferior gluteal
  9 Sciatic
 10 Posterior femoral cutaneous
 11 Pudendal
 12 Inferior hemorrhoidal

Inferior gluteal Can be located dorsally, deep to the caudofemoralis and gluteus maximus, both of which it supplies. It accompanies the inferior gluteal blood vessels.

Other branches from the lumbosacral plexus supply muscles in the hip region that are not supplied by the femoral, obturator, and gluteal nerves.

Sciatic A very large nerve, which you have already observed deep to the caudofemoralis, gluteus maximus, and biceps femoris. It supplies, through its branches, all of the dorsal femoral muscles and all of the crural muscles and intrinsic foot muscles. Its terminal branches are the **common peroneal** and **tibial nerves,** which you have observed. The common peroneal nerve divides, in turn, into a deep branch that supplies the ventral crural muscles and a superficial branch that supplies the lateral crural muscles; these branches continue into the foot. The tibial nerve supplies the dorsal crural muscles and continues into the foot.

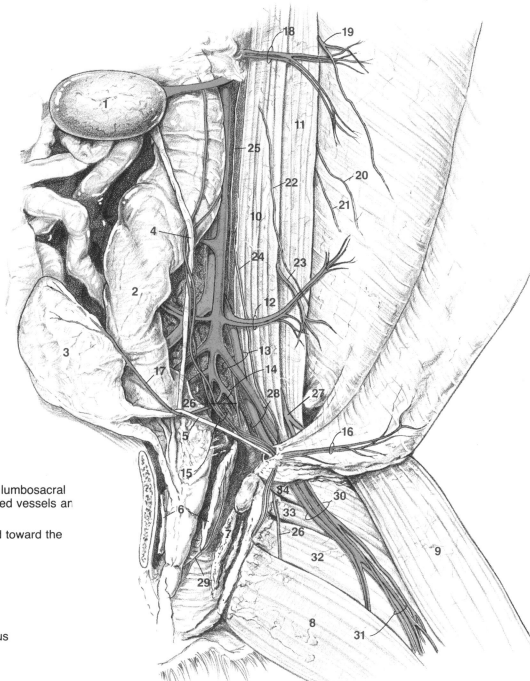

Figure 8-18
Some nerves of the lumbosacral
plexus with associated vessels and
muscles

1 Kidney (reflected toward the
   right side)
2 Rectum
3 Urinary bladder
4 Ureter
5 Vas deferens
6 Levator ani
7 Obturator internus
8 Gracilis
9 Sartorius
10 Psoas minor
11 Psoas major
12 Iliolumbar vessels
13 External iliac artery and
   common iliac vein
14 Internal iliac artery
15 Internal spermatic vessels
16 Inferior epigastric vessels
17 Umbilical artery
18 Adrenolumbar vessels
19 First lumbar nerve
20 Second lumbar nerve

21 Third lumbar nerve
22 Genitofemoral nerve, lateral
   branch
23 Lateral femoral cutaneous
   nerve
24 Genitofemoral nerve, medial
   branch
25 Sympathetic trunk
26 Obturator nerve

27 Femoral nerve
28 External iliac artery and vein
29 Pudendal nerve
30 Femoral vein, artery, and nerve
31 Greater saphenous vein,
   saphenous artery and nerve
32 Adductor femoris
33 Adductor longus
34 Pectineus

Before its division into common peroneal and tibial nerves, the sciatic gives off a cutaneous **sural nerve,** which supplies the skin of the dorsal and lateral leg and extends into the foot. It can be found along the popliteal border of the biceps femoris muscle, near the sural blood vessels, then coursing along the lateral surface of the gastrocnemius muscle, and further distad near the lesser saphenous vein. In the human, sural nerves branch from the common peroneal and tibial nerves.

Pudendal Supplies muscles and other tissues in the area of the external genitalia. It also gives off branches to the caudal end of the rectum and anal area. The nerve can be located laterad of the levator ani muscle, which can be detached from the innominate bone (see Fig. 8-18) and reflected.

Posterior femoral cutaneous Accompanies the inferior gluteal blood vessels and extends onto the thigh, where it can be located in the connective tissue on the lateral surface of the biceps femoris. It lies alongside the lesser saphenous vein for some distance and, along with the vein, passes deep at the caudal border of the caudofemoralis muscle. In addition to supplying the skin over the posterior surface of the thigh, this nerve sends branches to the anal area.

Inferior hemorrhoidal Follows the middle hemorrhoidal artery to the urethra, urinary bladder, and caudal end of the rectum.

In the pelvic region you will find small nerves that are given no specific names. These may supply skeletal muscles in the pelvic floor or may be part of an autonomic plexus of nerve fibers that supply major pelvic viscera.

INDEX